The Sub-Quantum Holoflux Theory of Consciousness

S. R. Joye, PhD

Published by the Viola Institute
Viola, California

2025

Material in this book has appeared in the author's previous books: *The Pribram-Bohm Holoflux Theory of Consciousness, The Electromagnetic Brain, Tuning the Mind, The Little Book of Consciousness,* and *The Little Book of the Holy Trinity.*

Printed in the United States of America

The Sub-Quantum Holoflux Theory of Consciousness

ISBN-13: 978-1-950761-25-8

Contents

OTHER BOOKS BY THE SAME AUTHOR

The Electromagnetic Brain: EM Field Theories on the Nature of Consciousness

Tuning the Mind: Geometries of Consciousness

Exploring the Noosphere: Teilhard de Chardin

Developing Supersensible Perception: Knowledge of the Higher Worlds through Entheogens, Prayer, and Nondual Awareness

The Little Book of Consciousness: Holonomic Brain Theory and the Implicate Order

Sub-Quantum Consciousness: A Geometry of Consciousness Based Upon the Work of Karl Pribram, David Bohm, and Pierre Teilhard de Chardin

PREFACE

This book is a rewrite of one of my previous books, *The Pribram-Bohm Holoflux Theory of Consciousness.* At the urging of numerous colleagues, I have made small refinements and improvements to the text and diagrams and have changed the title to *The Sub-Quantum Holoflux Theory of Consciousness* to reflect the fact that much of the theory was not directly developed by either Pribram or Bohm, but by myself during after years of dissertation research and subsequent years of writing and thinking about their work and how one might better integrate their ideas and those of others in order to understand and visualize a plausible, fact-supported geometry and cosmology of consciousness. My goal has always been to express such a model in lay terms, without the need to refer to higher mathematics, in order to maintain these ideas understandable for those readers seeking to better understand consciousness but not having a professional background in the fields of quantum physics, neurobiology, electrical engineering, and mathematics.

KEY TERMS AND DEFINITIONS

Explicate Order: One of the two primary dimensional orders in the universe, as posited by David Bohm. The explicate order encompasses spacetime and all manifestations within space- time. The other dimensional order Bohm termed the "Implicate Order."

Holosphere: A spherical boundary that has the diameter of one Planck length (3 x 10^{-35} meter). It is the event horizon between the explicate order (spacetime) and the implicate order (hidden dimensions).

Holoplenum: A continuous plenum of Planck length holospheres at the bottom of space.

Holoflux: Multidimensional holographic energy of information-consciousness. Holfolux in spacetime is projecting out of dimensions within Bohm's implicate order; holoflux is synonymous with the electromagnetic field in spacetime, resonating with multiple fields throughout other dimensions of the implicate order.

Implicate Order: One of two primary dimensional orders in the universe, as posited by David Bohm. The implicate order is a nonlocal, nontemporal domain that can be found below the Planck length. In holoflux theory it is equivalent to the frequency-phase domain in communications engineering. Bohm felt that the implicate is "somehow 'deeper' and more fundamental than the explicate" (Peat, 1987, p. 262).

Metaverse: The entirety of the conscious universe; an integrated network of multiple dimensions that make up the whole of reality. This includes, in addition to space and time, a minimum of at least an additional seven hidden dimensions (as predicted by string theory) as well as immunerable virtual reality dimensions open to creation, development, and exploration by consciousness.

Planck length: The smallest possible length in space, determined by Max Planck from the gravitational constant and the speed of light. The value of the Planck length is 3×10^{-35} meter.

Planck time: The smallest possible unit of time, determined by Max Planck from the gravitational constant and the speed of light. The value of Planck time is 5×10^{-44} sec.

Psychonauts: The smallest possible unit of time, determined by Max Planck from the gravitational

5

The day science begins to study non-physical phenomena, it will make more progress in one decade than in all the previous centuries of its existence.

– Nikola Tesla[1]

Currently, the terms “Dark Energy” and even ”Dark Matter” have surfaced as having to be measured and conceived in terms other than space and time. I conceive of both these “hidden” quantum and cosmological constructs as referring to a “potential reality” which lies behind the spacetime “experienced reality” within which we ordinarily navigate.

– Karl Pribram[2]
The Form Within, (2013), p. 526

Introduction: The Granularity of Spacetime

When I was a young physics student, I was intrigued by a lecture I heard by the California professor, Richard Feynman. He asked us to consider the following. If we were to take a hammer and try to strike an anvil with it, we could mathematically track the distance of the hammer from the anvil, at each moment of time. At some moment the hammer would reach half of the remaining distance to the anvil. You could plot the time of the next moment after that, when the hammer would again reach half of the distance remaining. If space were continuous, you could continue measuring the time at which the hammer managed to move closer to the anvil by halving the distance. But, if space were continuous, you would you never actually be able to strike the anvil because you could continue to move half the previous distance closer to the anvil. Thus, space is not continuous, there must be a final limit, so space must be granular, not continuous. This goes for time too, he said. Time is not continuous, but is granular. Then, having duly confused us students, Feynman would launch into his lecture on quantum physics.

Quantum means "discrete" in one sense, i.e. there is no continuity but a sort of "jump" (in fact the phrase "quantum jump" is often used). This "jump" is due to the bottom limits to space and time that were discovered to be the solution to solve a major problem that had plagued physicists for decades, i.e. the mathematics to model the relationship of temperature peaks to the peak wavelength of the emf in the infrared band (that we sense and term "heat").

Max Planck (known as "the father of Action") somehow intuited the solution that had escaped everyone else for decades. He had been working to solve the puzzle of the emission of radiation from heated objects, often called the problem of "black-body radiation."

All normal matter at temperatures above absolute zero emits electromagnetic radiation. It had long been noticed that charcoal could be seen to be giving off different colors that depended directly upon the temperature to which they had been heated. As the temperature of charcoal is increased, the visible emission began to be seen as a dull red, then as the temperature increased, it became more orange, then yellow, then green, and finally violet. Laboratory measurements had been made of the wavelength of light coming from glowing charcoal at different temperatures (fig. 1), but the calculations made (using "Classical theory," prior to quantum theory), shown in the dashed curve) were way off, and never matched the actual 5000 K measurements, shown in the solid line curves. In the figure can be seen the large discrepancy for a temperature of 5000 Kelvin.

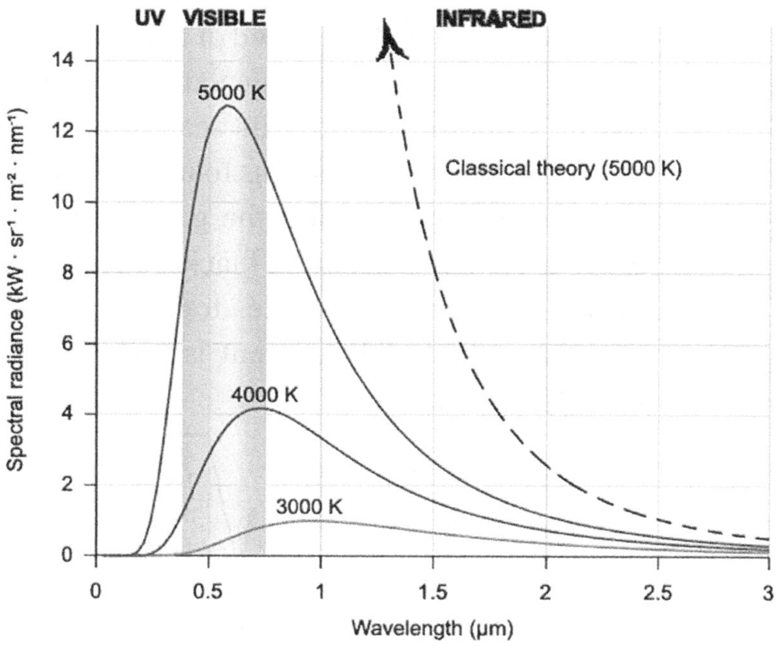

Fig. 1. Black-Body Radiation Curve Discrepancy[3]

Until Planck solved the puzzle, nobody had been able to accurately calculate the "black-body radiation curves" that would match laboratory measurements.

Planck's intuitive flash of insight was this:

1.) There can be no completely continuous Actions in space and time.

2.) There is both a top and bottom limit to the dimensions of time and space.

3.) Action must thus be "granular" (corollary: the universe is digital)

Knowing that the speed of light is invariant, Planck immediately sought to calculate the bottom limit of space, spending months in searching among various mathematical configurations to use the known constants of the speed of light and the gravitational constant in order to match exactly the black-body radiation curves. Eventually they all came together in his famous equation, which defines what is now called the Planck Length (fig. 2).

$$\ell_P = \sqrt{\frac{\hbar G}{c^3}}$$

ℓ_P = **The Planck Length**

$\hbar$ = ***Planck's constant***

G = ***Gravitational constant***

c = ***Speed of light***

Fig. 2. The Planck Length Equation

Because length divided by velocity results in time, he was also able to calculate the fundamental unit of time (fig. 3) as well as several other fundamental lower limits.

Calculation of the Planck length

$$\ell_P = \sqrt{\frac{\hbar G}{c^3}} = 1.616229 \text{ x } 10^{-35} \text{ meter}$$

Calculation of the Planck time

$$t_P = \frac{\ell_P}{c} = \sqrt{\frac{\hbar G}{c^5}} = 5.391247 \text{ x } 10^{-44} \text{ seconds}$$

Fig. 3. Calculation of the Planck length and the Planck time constants

These then are the lower limits to our space and time dimensions:

Planck length constant ➔ 1.616255×10^{-35} Meters

Planck time constant ➔ 5.391247×10^{-44} Seconds

Planck mass constant ➔ 2.176434×10^{-8} Kg

Planck temperature constant ➔ 1.416784×10^{32} Kelvin

Note that at the smallest dimension, the Planck length, the temperature reaches the maximum possible in spacetime , the Planck temperature constant. That is because at smaller dimensions, the radii of the spinning energy "strings" is smaller, and thus the frequencies of the emitted radiation are higher. Higher frequencies emit radiation at higher energy levels.

Henceforth, using these constants in the mathematical model made the curve calculations match the observed curve exactly. A year after his discovery, Max Plank began using the term "quanta" (Latin: "how much" or "amount") to describe these values, and some years later, his colleagues began to call these these the "Planck constants." In 1918 Planck was awarded the Nobel Prize for his discovery of these constants.

The Planck time constant of 5.391247×10^{-44} seconds is the smallest unit of time possible in our spacetime continuum. While humans are familiar with the duration of "one human second" of time, conventionally measured with our own time keeping clock devices, it is difficult to imagine much smaller time units. One of the fastest contemporary Intel microprocessor chips runs at 4.7 GHz, which is 4.7×10^{9} . However within this "one human second" there are 10^{9} "processor ticks" and 10^{44} "Planck time ticks" per second, which is truly mind-boggling!

	Scientific Notation	Seconds, in Standard Notation
Standard "second"	10^{0}	1 (note that this sets the scale of human-perceived time to approximately equal the average human heartbeat)
Early computer processor (Z80 chip)	10^{-5}	0.00001
Intel microprocessor (Intel Core i5 chip)	10^{-9}	0.000000001
Planck time	10^{-44}	0.00000000000000000000000000000000000000000001

Comparison of time units: Scientific notation and Standard notation.

Because time does not exist below this lowest granular division of time, Planck time, there is a quantum "jump" as time flows, a jump from one tick to the next. But can we imagine what occurs between the two ticks? In between these ticks, time does not exist. This is what metaphysicians would call the void, the timeless eternal, the transcendent. In the cosmological physics of David Bohm, it is the implicate order, the ground of being *into which* all information folds inward from our external spacetime , and from *out of which* all things are unfolded and projected into spacetime by the same implicate order.

To Bohm, all particles that exist within our cosmos of spacetime must be considered to be actually projections out of a higher dimensional order:

> We may thus regard each of the 'particles' constituting a system as a projection of a 'higher-dimensional' reality, rather than as a separate particle, existing together with all the others in a common three-dimensional space. For example, in the experiment of Einstein, Podolsky and Rosen, each of two atoms that initially combine to form a single molecule are to be fully regarded as three-dimensional projections of a six-dimensional reality.[4]

Pribram and Bohm spoke about a holomovement occurring outside of the four dimensions of space and time. Pribram did not like the use of the word "movement" in this naming convention, and recommended to Bohm that they use the term "holoflux," which is not a movement in time, but more of an evolution of information within the frequency domains of the implicate order. In fact one of the attributes of the implicate order, as Bohm understood it, would be a continuous *enfolding* of information from the spacetime domain, back into the hidden dimensions of the implicate order, where somehow the total information coming from spacetime for that one "tick" of time would be immediately superimposed over all of the pre-existing information stored in the implicate order (Akasha, according to Bucke, Steiner, and Laszlo). At the very next quantum "tick" of time, the cosmos would be reprojected into spacetime from information holoflux within the implicate order. The cycle can truly be seen to be a digital process, with the cosmos being created and recreated with each tick of the Planck-time clock. Bohm says:

> What follows from all this is that basically the implicate order has to be considered as a process of enfoldment and unfoldment in a higher-dimensional space. Only under certain conditions can this be simplified as a process of enfoldment and unfoldment in three dimensions.[5]

Bohm goes on to describe this projected information holoflux as it appears within the constricted spacetime dimensions:

> The electromagnetic field, which is the ground of the holographic image, obeys the laws of the quantum theory, and when these are properly applied to the field it is found that this, too, is actually a multidimensional reality which can only under certain conditions be simplified as a three-dimensional reality. Quite generally, then, the implicate order has to be extended into a multidimensional reality. In principle this reality is one unbroken whole, including the entire universe with all it's 'fields' and 'particles.'[6]

The Sub-Quantum Holoflux Topology

The psychophysical model of consciousness which is developed here is presented in fig. 4, where the basic theory is diagrammed as consciousness transforming between nonlocal and local regions of experience and information.

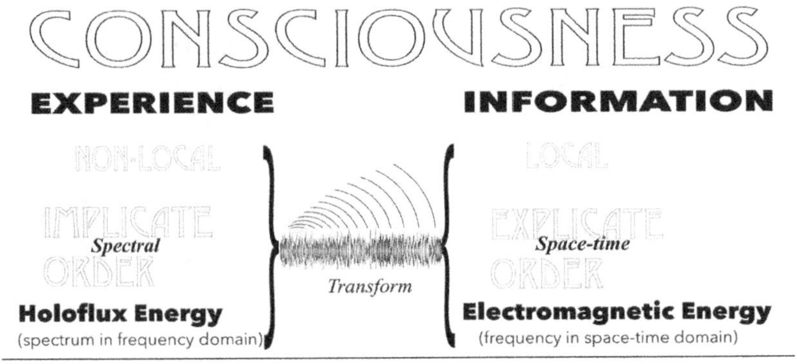

Fig. 4. The Pribram-Bohm Holoflux Model

The sub-quantum holoflux hypothesis regards consciousness as a cybernetic energy process, a holoflux transforming between two orders of being in "an undivided flowing movement without borders" (Bohm, 1980, p. 172). To the left in the diagram of Figure 1, consciousness is expressed as a spectrum of holoflux energy in Bohm's implicate order. This holoflux energy resonates with electromagnetic energy of the same frequencies to the right in the diagram, in the spacetime region, or explicate order (Bohm, 1980, p. 159).

Viewed from left to right, the diagram reveals a spectrum of holoflux energy in the transcendental implicate order transforming and translated into "things" and "events" in local spacetime , and conversely, viewing the diagram from right to left, information generated by "things" and "events" interacting throughout spacetime is seen to be transforming (folding) back into the implicate order. The process is described as a continuous cybernetic cycle, perhaps occurring at a regular clock-rate.

Superposition of Consciousness
Cybernetics and the Fourier Transform

Common experience would suggest that all awareness is consciousness *of* something, such as the experience of a sound, of an image, of a sensation, of an emotion, of an interior verbal thought. These experiences seem to be superpositioned, they often occur at what seem to be the same perceptual moment. Yet each simultaneous stream of experience remains distinct, somehow integrated with all the others. But what is it that is "looking at" this stream of experiences? It is as if there is some meta-consciousness that is more than the sum of each of these individual "experience streams," per se, but rather some other, more comprehensive level, some panoramic perspective that is able to embrace and comprehend them all, and which has the amazing ability to *fine tune* its own selected focus upon *one or more* of these streams of awareness while simultaneously dampening and filtering out the many others.

In signal analysis this stream phenomenon is explained by the *superposition principle*, formalized in 1822 by the French mathematician Jean-Baptiste Fourier, who developed the mathematics of what is now called "Fourier analysis" during his search for a mathematical relationship between spacetime and frequency (Feynman, Leighton, & Sands, 1964, p. 286). Because signals are more readily superpositioned and manipulated (filtered, amplified, etc.) within the *frequency domain* than in the *time domain*, the Fourier transform equations have become primary and ubiquitous mathematical tools in physics and engineering for analyzing, synthesizing, and transmitting signals between two domains:

1. a "spacetime domain (t_d)," and

2. a "frequency domain (f_d)."

Much of electrical engineering circuit design is done within the frequency domain, and only subsequently implemented with time domain components, as described here by Francis F. Kuo, chief electrical engineer at the original Bell Telephone Laboratory from which emerged the transistor, the laser, and radio astronomy, information theory, and the Unix operating system. In his textbook on *Network Analysis and Synthesis*, Kuo (1962) states:

> We see that in the **time domain** (i.e., where the independent variable is **t**) the voltage–current relationships are given in terms of differential

> equations. On the other hand, in the complex **frequency domain**, the voltage–current relationships for the elements are expressed in **algebraic** equations. Algebraic equations are, in most cases, more easily solved than differential equations. Herein lies the **raison d'être** for describing signals and networks in the frequency domain as well as in the time domain. (p. 13)

Norbert Wiener (1948) coined the term "cybernetics" from the Greek κυβερνήτης— "steersman, governor, pilot, or rudder" (p. 11)—during his own work at the same Bell Telephone Laboratory as Kuo, and made use of Fourier's transform to model and analyze brain waves in the frequency domain, where he discovered clear evidence of "self-organization of electroencephalograms or brain waves" (p. 181). Using Fourier analysis, an approach which later became of great interest to Bohm, Wiener (1948) was able to detect uniquely narrow frequency ranges, centered within different spatial locations on the cortex, that repeatedly exhibited auto- correlation (p. 191). Regions on the cortex were identified where specific ranges of frequencies were found to coalesce toward intermediate frequencies, seeming both to attract and to strengthen one another, exhibiting *resonance* or "self tuning" to amplify and consolidate signals into narrowly specific ranges in the frequency domain f_d (p. 198). His research led Wiener to conjecture that the *infrared band* of electromagnetic flux may be the loci of "self–organizing systems":

> We thus see that a nonlinear interaction causing the attraction of frequency can generate a self-organizing system, as it does in the case of the brain waves we have discussed. This possibility of self-organization is by no means limited to the very low frequency of these two phenomena. Consider self–organizing systems at the frequency level, say, of infrared light. (p. 202)

Three years after Wiener's (1948) publication of *Cybernetics*, David Bohm (1951) stressed the importance of Fourier's equations on the first page of his well-received 646-page textbook, *Quantum Theory*, where he encouraged a familiarity with Fourier analysis for an ontological understanding of quantum phenomena:

> It seems impossible to develop quantum concepts extensively without Fourier analysis. It is, therefore, presupposed that the reader is moderately familiar with Fourier analysis. (p. 1)

For purposes of this discussion, the basic understanding of Fourier analysis is simply that *frequency vibrations* manifest within two distinct dimensions or domains: a spacetime domain and a frequency domain. Until recently, physicists have focused exclusively within spacetime to conduct their research, considering only space and time as having any "reality" and considering the ontological reality of the frequency domain, if at all, in the same vague category as the domain of mathematics itself (i.e., in some unspecified transcendent dimension). Whether there might somehow exist a "real" dimension *outside of* spacetime , or *beyond* spacetime has generally been beyond the purview of the physical sciences. Yet the experienced reality of a region of consciousness beyond spacetime is supported by the vast body of first–hand reports generated by religious, mystical, or near-death experiences. In an approach to such experiences, William James (1909), the "father of American psychology," writes:

> The further limits of our being plunge, it seems to me, into an altogether other dimension of existence from the sensible and merely "understandable" world. Name it the mystical region, or the supernatural region, whichever you choose. So far as our ideal impulses originate in this region (and most of them do originate in it, for we find them possessing us in a way for which we cannot articulately account), we belong to it in a more intimate sense than that in which we belong to the visible world, for we belong in the most intimate sense wherever our ideals belong. (p. 318)

Fourier's transform equations (fig. 5) between the two domains of time (t_d) and frequency (f_d) are more than simply mathematical equations, written down as functions in the abstract symbolic language of calculus (Stein & Shakarchi, 2003, pp. 134–36).

$$f(t) = \int_{-\infty}^{+\infty} X(F) e^{j2\pi Ft} dF \qquad f(F) = \int_{-\infty}^{+\infty} x(t) e^{-j2\pi Ft} dt$$

Fourier integral transform of a continuous frequency function into the time domain (t_d).

Fourier integral transform of a continuous time function into the frequency domain (F_d).

Fig. 5. The Fourier transform and inverse transform

These two expressions indicate that any function in the timespace domain can be transformed into and expressed equivalently as an infinite series of frequency spectra functions in the frequency domain. The transformation is also possible in the opposite direction, such that any arbitrary function in the frequency domain, can be transformed into and expressed by an infinite series of time functions. The two domains mirror one another.

Beyond purely mathematical considerations, the equations can be taken as models of an actual cosmic process (i.e., much as Newton's Law model the phenomenon of gravity) and they can be understood as mirroring the cosmos in mathematical terms. The model of consciousness presented in this paper proposes that there is indeed an ontological reality to this other region, and that this region is synonymous with Bohm's "implicate order," Pribram's "holonomic frequency domain."

Pribram's Holonomic Mind/Brain Theory

The neurosurgeon Karl Pribram (1971) was one of the first to articulate the idea that the Fourier transform might play a role in brain/mind neurophysics. Pribram (1990) spent decades performing laboratory research to gather experimental data in an effort to solve two problems:

(a) to identify the location and mechanism of memory storage (the *engram*), and (b) to discover the cognitive mechanism behind visual perception. Pribram (2013) arrived at the conclusion that the data revealed evidence of Fourier signal transformations of visual signals from the rods and cones of the eyes, and that these Fourier patterns could be detected in spatial Fourier patterns over wide areas of the brain, as fields within the fine-fibered dendritic networks of the cerebral cortex (p. 82).

In the mid-1960s, Pribram was inspired by reports of the first optical holograms, and the empirical evidence that holograms could store, retrieve, and process vast quantities of information using resonant photons in high frequency beams. Ten years later, Pribram (1971) published *Languages of the Brain*, in which he detailed his new theory, the holonomic brain/mind theory, based upon evidence of the Fourier transform playing a key role in the mind/brain process. The theory he put forth proposed that the cognitive sensory processes of memory, sight, hearing, and consciousness in general, may all operate holographically, in a transformational process of information-coded-energies flowing back and forth between space– time and the frequency domain via a Fourier transform mechanism.

Pribram's (1971) theory was radical and controversial, challenging two prominent paradigms of modern neurophysical research: (a) the belief that consciousness is an epiphenomenon produced by electrical sparks among synaptic–clefts throughout the wiring of neurons the brain, and (b) the belief that somewhere in the physical brain, *engrams* of memory are stored, and will be eventually found. Pribram (1971) relates a story of a conversation he had at the time, while climbing with colleagues on a hike in Colorado just prior to attending a neuroscience conference in Boulder:

> We had climbed high into the Rocky Mountains. Coming to rest on a desolate crag, a long meditative silence was suddenly broken by a query from Campbell: "Karl, do you really believe it's a Fourier?" I hesitated, and then replied, "No Fergus, that would be too easy, don't you agree?" Campbell sat silently awhile, then said, "You are right, it's probably not that easy. So what are you going to say tomorrow down there?" I replied, this time without hesitation, "That the transform is a Fourier, of course." Campbell smiled and chortled, "Good for you! So am I." (p. xvii)

Pribram's (1971) hypothesis was strengthened through a growing appreciation of holography as frequency-superpositioned electromagnetic wave interference (p. 142). Pribram called his approach "the holonomic brain theory," and postulated the importance of the *frequency domain* in future research:

> Essentially, the theory reads that the brain at one stage of processing performs its analyses in the frequency domain . . . a solid body of evidence has accumulated that the auditory, somatosensory, motor, and visual systems of the brain do in fact process, at one or several stages, input from the senses in the frequency domain. (Pribram, 1982, p. 29)

In Pribram's (1990) theory, a pure frequency domain links with the neuronal tissue of the brain through modulating fields of flux within the fine-fibered dendritic webs of the cerebral cortex regions. His paradigm was reinforced at a San Francisco conference during a lecture given by the physicist Geoffrey F. Chew, the head of the UC Berkeley physics department and a former student of Enrico Fermi. Chew presented a conceptual diagram of the Fourier transform process (fig. 6), which perfectly encapsulated what Pribram had by then become familiar with, the Fourier transform (Pribram, 2004a, p. 230). As shown in the figure, the spectral (frequency) domain, located at the left of the diagram, is directly linked to the spacetime domain, depicted at the right, bridged by the Fourier transform, operating at the sub-atomic levels predicted by Planck's constant.

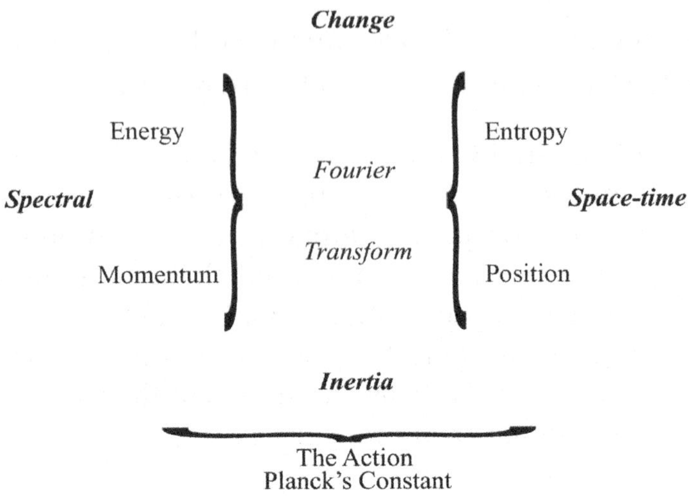

Fig. 6. The Dirac Fourier transform diagram. Source: Adapted from Pribram (2004b)

Pribram (2013) asked Chew where he had obtained the diagram, and was told that he had been given the diagram by his colleague at Berkeley, the physicist Henry Stapp, who himself said he had been given it directly from the British theoretical physicist Paul Dirac (1902–1984), one of the original founders of quantum mechanics. Whatever the origin of the figure, Pribram chose to include the diagram in several future papers. In "Consciousness Reassessed," Pribram's (2004b) caption to the figure reads, "The Fourier Transform as the Mediator between Spectral and Spacetime " (p. 8).

In the diagram, the spectral domain is shown at the left and spacetime to the right, with the Fourier transform between them. The diagram became foundational to Pribram's understanding. It presents a two–way Fourier transform, operational at the boundary between the two domains, located at an event horizon termed in the diagram, "The Action: Planck's Constant." It is this process of turbulent transformation at the event–horizon that David Bohm and Basil Hiley (1993) termed holomovement or holoflux (p. 382).

The Limits of Space
From the Edge of the Universe to Planck's Constant

The sub-quantum holoflux hypothesis holds that the dimensions of space are finite and that space exhibits a limited domain in a quantifiable range. This is consistent with the physics of string theory or M–theory, [quote on dimensions moved to first chapter]

Carr (2005) uses the alchemical image of the ouroboros (fig. 7) to illustrate his GUT theory (Grand Unified Theory) in comparing major scale-dependent structural levels of the physical world: "The significance of the head meeting the tail is that the entire Universe was once compressed to a point of infinite density (or, more strictly, the Planck density)" (p. 13). This archetypal figure implies the interconnectedness of the entire universal process in time and space, presenting a cybernetic feedback loop operational at every scale. Mystics have intuited this ouroboric process symbolized in the images of a snake swallowing its own tail (the image has been found as early as the 14th century BCE in the tomb of Tutankhamun) and it is frequently used to symbolize cybernetic feedback in control and communication theory (Wiener, 1948).

While Wiener coined the term *cybernetics*, communication engineers would more commonly see this as metaphor for the "feedback loop," used everywhere in electronic circuit design.

Fig. 7. Alchemical Ouroboros (Pelekanos, 1478)

Stretching out this circular cosmic ouroboric serpent from head to tail, one can create an axis of scales that encompasses all of space. In fig. 8 such a scale is drawn starting with the currently estimated diameter of the universe itself at 10^{+25} m, and descending logarithmically down to the Planck length limit at 10^{-35} m. The axis thus spans a total range of 10^{+60} (60 jumps by the power of 10). The Pribram–Bohm hypothesis holds that there, at the very bottom of the linear scale (fig. 8), is to be found the transition bounding the explicate order and the implicate order . Here, at the bottom bound of the spatial scale, space reaches its *end*, according to modern physics; but it also marks the *entry point into* Bohm's "implicate order," what Pribram terms the "frequency domain" (Bohm, 1980; Pribram, 2013).

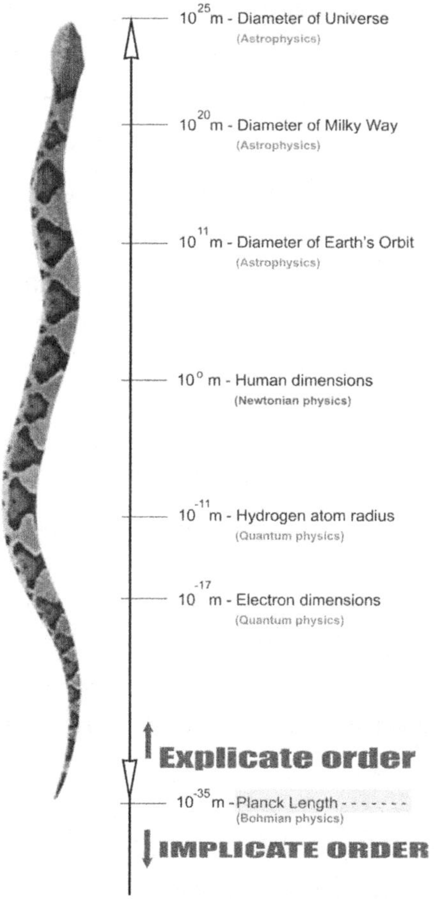

Fig. 8. Scales of dimensional space and the explicate/implicate boundary

The implications of this topology are profound. Imagine moving inwardly, from any position in the universe, moving into a spherical bubble, shrinking ever smaller in scale while moving ever closer to the center at the bottom of the spatial scale, following the radial axis inward, ever shrinking downward, and then abruptly reaching the end of the line at the Planck length limit of space, the locus of a spherical shell 10^{-35} meters in diameter, below which space has no meaning. Here a boundary has been reached, an event horizon between space and the implicate order. To understand this, one must realize that the classical Cartesian assumption that space is continuous is *wrong*; there *is* indeed a bottom to space, at least according to physics, below which space no longer has meaning. Here there is a discontinuity, as David Bohm and F. David Peat (1987) explain in describing the granularity of space:

> What of the order between two points in space? The Cartesian order holds that space is continuous. Between any two points, no matter how close they lie, occur an infinite of other points. Between any two neighboring points in this infinity lies another infinity and so on. This notion of continuity is not compatible with the order of quantum theory
> Thus the physicist

John Wheeler has suggested that, at very short distances, continuous space begins to break up into a foam-like structure. Thus the "order between" two points moves from the order of continuity to an order of a discontinuous foam. (pp. 311–312)

Pribram's Spectral Flux and the Implicate Order

In 1979, Karl Pribram, at that time a Stanford professor, attended a conference in Cordoba, Spain, where he met David Bohm, a professor of theoretical physics at London University (Cazenave, 1984). During the conference, Pribram (2013) soon realized that David Bohm's model of the implicate order and its projection, or extrusion into space–time, could be seen as entirely compatible with his own holonomic mind/brain theory. Thus began 20 years of correspondence and dialog between David Bohm and Karl Pribram, and the two soon became personal friends.

Pribram (2013) saw in Bohm's theories how the frequency domain flux might be seen to unfold into explicate domain waves of encoded information via the Fourier transform, and he appreciated Bohm's description of how information from the explicate may fold back into the implicate in a bi-directional process. Even more intriguing was Bohm's belief that, "the basic relationship of quantum theory and consciousness is that they have the implicate order in common." (Bohm & Hiley, 1993, pp. 381–382)

Pribram was equally impressed with Bohm's explanation of nonlocality, a major mystery in quantum physics, which Bohm explains as fundamental to the process of folding and unfolding between explicate and implicate orders, allowing for full superpositioned cohesion of frequency information within the implicate order, and even providing a plausible mechanism for Sheldrake's theories of morphogenetic fields and morphic resonance:

> The implicate order can be thought of as a ground beyond time, a totality, out of which each moment is projected into the explicate order. For every moment that is projected out into the explicate there would be another movement in which that moment would be injected or "introjected" back into the implicate order. If you have a large number of repetitions of this process, you'll start to build up a fairly constant component to this series of projection and injection. That is, a fixed disposition would become established. The point is that, via this process, past forms would tend to be repeated or replicated in the present, and that is very similar to what Sheldrake calls a morphogenetic field and morphic resonance. Moreover, such a field would not be located anywhere. When it projects back into the totality (the implicate order), since no space and time are relevant there, all things of a similar nature might get connected together or resonate in totality. When the explicate order enfolds into the implicate order, which does not have any space, all places and all times are, we might say, merged, so that what happens in one place will interpenetrate what happens in another place. (Bohm & Weber, 1982, pp. 35–36)

Bohm's topology is both supported and extended by Pribram's contention, supported by the diagram handed down from Dirac, that the boundary or event horizon between the two domains, where the action occurs, is at the Planck length, precisely where, as Pribram tells us here, spectral density in-formation translates into spacetime ex-formation.

> Matter can be seen as an "ex-formation," an externalized (extruded, palpable, compacted) form of flux. By contrast, thinking and its communication (minding) are the consequence of an internalized (neg-entropic) forming of flux, its "in-formation." My claim is that the basis function from which both matter and mind are "formed" is flux (measured as spectral density). (Pribram, 2004b, p. 13)

This flux or spectral density is for Pribram real, in the same sense that spacetime is considered to be real, but this flux is *outside of* or *beyond* space–time. It is in this sense that Pribram made the conceptual leap from considering the Fourier transform as simply a tool of mathematical calculation, to a dawning realization that the reality of the transform implies the ontological *reality* of a domain *outside of space–time*, a transcendent yet ontologically real domain where energy as flux is "measured as spectral density."

Dirac's original diagram can now be extended to include Bohm's two regions of the whole, the implicate order and the explicate order. Figure 9 depicts this expanded diagram.

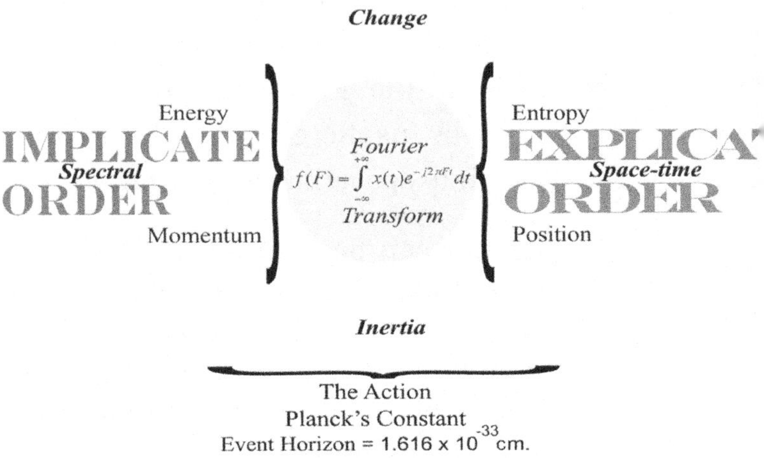

Fig. 9. Dirac Fourier diagram with David Bohm's topology

An anthropomorphic view of the Pribram's diagram can be seen in fig. 10, where an iris-like lens peering out from the implicate order is maintaining a focus upon and/or projecting a holonomic universe within the explicate order of space–time. This mirrors Karl Pribram's (1991) conceptualization of a lens between the two domains, expressed here in *Brain and Perception*:

> These two domains characterize the input to and output from a lens that performs a Fourier transform. On one side of the transform lies the spacetime order we ordinarily perceive. On the other side lies a distributed enfolded holographic–like order referred to as the frequency or spectral domain. (p. 70)

CONSCIOUSNESS

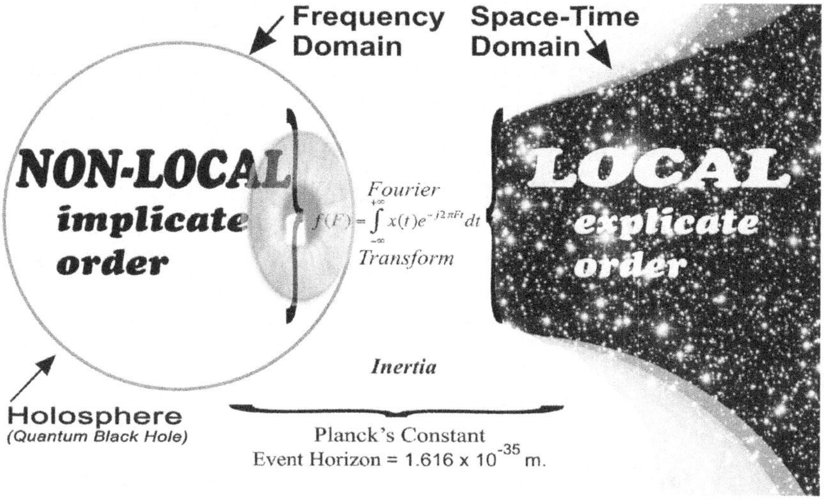

Fig. 10. Topology of consciousness

Note that the image of an iris in the diagram appears at the edge of the event horizon of a quantum black hole, or implicate order holosphere. The iris symbolizes consciousness looking *out* from the implicate order *into* spacetime via a Fourier transform lensing process. This approach to a topology of consciousness as something that is looking out and seeing itself is supported here by the mathematician G. Spencer-Brown (1972) in *Laws of Form*:

> Now the physicist himself, who describes all this, is, in his own account, constructed of it. He is, in short, made of a conglomeration of the very particulars he describes, no more, no less, bound together by and obeying such general laws as he himself has managed to find and record. Thus we cannot escape the fact

> that the world we know is constructed in order (and thus in such a way as to be able) to see itself. This is indeed amazing. Not so much in view of what it sees, although this may appear fantastic enough, but in respect of the fact that it can see at all. But in order *to do so, evidently it must first cut itself up into at least one state which sees, and at least one other state which is seen. In this condition it will always partially elude itself. (p. 105)*

Cosmology and Bohm's Implicate Order

In 1980 Bohm published *Wholeness and the Implicate Order*, and in a section in which he discusses the cosmology of the implicate order, he puts forth a solution to the problem of "zero-point" energy by regarding the Planck length as the shortest wavelength possible:

> If one were to add up the energies of all the "wave-particle" modes of excitation in any region of space, the result would be infinite, because an infinite number of wavelengths is present. However, there is good reason to suppose that one need not keep on adding the energies corresponding to shorter and shorter wavelengths. There may be a certain shortest possible wavelength, so that the total number of modes of excitation, and therefore the energy, would be finite. When this length is

> estimated it turns out to be about 10^{-35} m. *(Bohm, 1980, p. 190)*

Bohm (1980) brings up the school of Parmenides and Zeno, which held that all of space is actually a plenum, and he points out that as recently as the last century this same theory was presented in the widely accepted hypothesis of the *ether* (p. 191). Bohm describes how there is a "holomovement" (p. 151) in this immense sea of "zero-point energy" (p. 190) to be understood as a "undivided flowing movement without borders" (p. 172) and he goes on to state:

> It is being suggested here, then, that what we perceive through the senses as empty space is actually the plenum, which is the ground for the existence of everything, including ourselves. The things that appear to our senses are derivative forms and their true meaning can be seen only when we consider the plenum, in which they are generated and sustained, and into which they must ultimately vanish. (p. 192)

In the Pribram–Bohm cosmology then, the interface or boundary between the spacetime explicate domain and the nonlocal, nontemporal implicate domain can be viewed topologically as a holoplenum of holospheres (Figure 8). Here can be found an answer to the "hard problem of consciousness" posed by Chalmers (1995), for it is from *within* each holosphere that consciousness is "peering out" into and "projecting" the spacetime explicate, and here Bohm (1980) summarizes his cosmological essay by proposing that "consciousness is to be comprehended in terms of the implicate order, along with reality as a whole" (p. 196) and stating unequivocally that "the implicate order is also its primary and immediate actuality" (p. 197).

How Spacetime Projects from the Holoplenum

To summarize thus far, the Pribram–Bohm model envisions the existence of a nonlocal, transcendent mode of consciousness out of which projects the spatial cosmos holographically from an infinite plenum or matrix of quantum black holes, each at the limiting Planck length diameter of 10^{-35} m, located at the very bottom of space, at the very center of every three-dimensional spatial coordinate. According to quantum theory, within the implicate order, below the Planck length, space and time do not exist, as Bohm (1980) describes here:

> We come to a certain length at which the measurement of space and time becomes totally indefinable. Beyond this, the whole notion of space and time as we know it would fade out, into something that is at present unspecifiable When this length is estimated it turns out to be about 10^{-33} cm. This is much shorter than anything thus far probed in physical experiments (which have gone down to about 10^{-17} cm or so.*) (p. 190–191)*

This leads to a new vision of the metaverse, a model in which the cosmos can be visualized as projecting outwardly from a holoplenum, from each Planck holosphere (quantum black hole) at the bottom of space, everywhere. In such a cosmology, the Big Bang theory would need to be revised to encompass more than simply a single point at a single time. The holoflux theory offers the prospect of multiple "Big Bangs" emanating from every individual Planck holosphere, perhaps a continuous recurrence cycling at Planck time (10^{-44} s) quantum ticks. Like images arising from pixels on a two-dimensional LCD (liquid crystal display) screen, the holoplenum projects a three-dimensional cosmos in all its glory. Yet during the process not only is radiant energy projecting *outwardly from* these quantum holospheres, but there is a simultaneous torrent of information, perhaps encoded in the form of gravitationally modulated energy flowing *inwardly into* the implicate order at the center of each and every point in space.

One can only imagine the transitional region between the explicate and implicate order as a frothing, turbulent, resonant event-horizon that bounds spacetime and the transcendent implicate order. Information flowing inward eventually leaves spacetime entropically and flows *into* the network of quantum black holes, the holoplenum of holospheres. Information flowing into this region of infinite centers becomes immediately nonlocal, superpositioned, omni-intersecting. The holoplenum of holospheres provides a vast 3D projection field into which spacetime is projected by implicate order holoflux.

But in addition to this spatial topology, an examination of the transition between time (in the explicate order) and timelessness (in the implicate order) is in order. How might time communicate with the timeless? What happens at the event-horizon of time?

The Digital Universe: Laminated Spacetime

At the heart of quantum theory is the conviction that neither time nor space are continuous and it has been concluded that both are granular at their lowest scale (Smolin, 2012). Having thus far established the granularity of space at the Planck length of 10^{-35} m, what then might be the granularity of time? One approach has been provided by the theoretical physicist John Archibald Wheeler (the originator of the term "black hole"), who, in considering a topology of information in the cosmos, noted that the highest possible clock rate must be limited by the Planck constant in general, and specifically by the *Planck time* value, below which time can have no meaning according to modern physics (Wheeler, 1990). The *Planck time* (t_P) is the time it would take a single photon travelling at the speed of light to cross a distance equal to one Planck length; the Planck length has been determined by the National Institute of Standards and Technology (n.d.) to have a value of 5.39116×10^{-44} s. Wheeler (1990) here describes the granular activity within time and space:

> Space––pure, empty, energy–free space––all the time and everywhere experiences so-called quantum fluctuations at a fantastically small scale of time, of the order of 10^{-44} seconds. During these quantum fluctuations, pairs of particles appear for an instant from the emptiness of space. (p. 222)

Given the limit of the smallest interval of time being 10^{-44} seconds, one can calculate the maximum clock–speed of the universe by determining how many Planck time intervals are possible within one human second:

$$(1 \text{ sec} / 10^{-44} \text{ sec}) = 10^{+44}$$

Thus it has been proposed that if the universe is found to operate digitally, it would be cycling at its highest possible clock rate, switching between the explicate order and the implicate order cyclically at a rate of 10^{44} times per second at a bandwidth of 10^{44} Hz. This model has been developed in *The Theory of Laminated Spacetime* by Dewey (1985). The theory accords well with the Pribram–Bohm hypothesis process, and describes a process of energy flux extruding into spacetime as a series of quantum shells of information–encoded energy moving out in quantum jumps at the speed of light, with each quantum shell or brane separated from the next by a spatial gap equal to the Planck length, as described here:

> I believe the universe to be composed of nothing but shells of electromagnetic particles which the theory of Laminated Spacetime describes as laminae of spacetime . (p. 95)

Holograms and the Holoplenum

To better understand the projection of the spacetime cosmos from the underlying holoplenum, the distinction between a hologram and a three dimensional holoplenum projection must be drawn. A hologram is an optical image stored and recreated by a single laser (i.e., LASER: Light Amplitude Stimulation of Electromagnetic Radiation), a beam of coherent electromagnetic radiation of a single frequency (Pribram, 1971, pp. 145–152). Human technology greatly simplifies the creation and replay of a hologram by: (a) using a single beam of coherent photon energy (a *single frequency*), and (b) illuminating the object from only two distinct fixed points in space. An interference pattern forms from the beams of intersecting light impinging upon the three–dimensional planes of the object from different angles, causing complex shadows which are detected and recorded on the flat detector plane behind the object (Pribram, 1971, p. 142).

In nature, however, the situation is vastly richer in complexity: electromagnetic flux interactions in spacetime form a highly complex three–dimensional matrix of intersecting shells of every spacetime radiation frequency band conceivable, impinging from an infinity of directions. In contrast with the human generated holograms of a single frequency taken from two fixed points, each point in spacetime actually intercepts the *entire frequency flux spectrum* of the cosmos, which is being continuously captured by and feeding the event horizon entropy of the quantum black hole located at its geometric central "point" (Wheeler, 1998, pp. 313–314).

The omnidirectional interactions manifesting among electromagnetic waves in the dimensional range of the explicate order are mirrored in the spectral frequencies within the implicate order. There is a resonance between implicate and explicate mediated by mathematical relationships such as the Fourier transforms and Gabor functions. This process between the explicate and implicate order is skillfully articulated here by David Bohm:

> The implicate order can be thought of as a ground beyond time, a totality, out of which each moment is projected into the explicate order. For every moment that is projected out into the explicate there would be another movement in which that moment would be injected or "introjected" back into the implicate order. If you have a large number of repetitions of this process, you'll start to build up a fairly constant component to this series of projection and injection. That is, a fixed disposition would become established. The point is that, via this process, past forms would tend to be repeated or replicated in the present, and that is very similar to what Sheldrake calls a morphogenetic field and morphic resonance. Moreover, such a field would not be located anywhere. When it projects back into the totality (the implicate

> order), since no space and time are relevant there, all things of a similar nature might get connected together or resonate in totality. When the explicate order enfolds into the implicate order, which does not have any space, all places and all times are, we might say, merged, so that what happens in one place will interpenetrate what happens in another place. (1981, pp. 35–36)

This resonance between the implicate and the explicate orders nudges into existence a cosmic holonomic metaverse of galaxies, butterflies, and zebras (Sheldrake, 1981).

Geocentric Topology of the Holonomic Metaverse

The topology of the holonomic metaverse is congruent with cosmological intuitions of numerous classical thinkers. In the 4th century BCE, a model proposed by Plato, and further developed by his student Aristotle, consisted of system of numerous crystalline shells rotating about a central sphere located at the center of the Earth (C. G. Fraser, 2006, p. 14). By the 2nd century CE, this geocentric model had become codified by the Alexandrian astronomer Claudius Ptolemy in his cosmological model of concentric spheres (fig. 11).

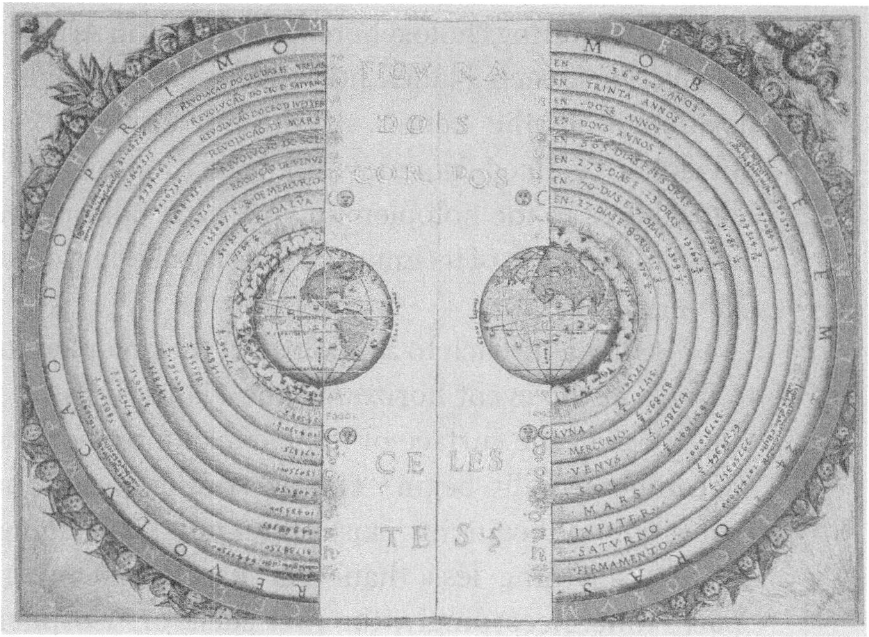

Fig. 11. The Ptolemaic Geocentric conception of the Universe. Source: Velho (1568)

In the 3rd century BCE Aristotle (2004) held that the reason an apple falls to the ground is because it seeks its natural place at the center of the universe, and he set forth a geocentric model based upon the following three propositions:

1. The Earth is positioned at the center of the universe.

2. The Earth is fixed (nonmoving) in relation to the rest of the universe.

3. The Earth is special and unique compared to all other heavenly bodies.

Substituting "holosphere" for "Earth" in Aristotle's propositions, each Planck holosphere can be taken as positioned at the center of the its universe; each holosphere is fixed (nonmoving) in relation to all other holospheres in the holoplenum, and each holosphere is "special" by virtue of its unique Hilbert space coordinates (Young, 1988).

A first approach to a topology of the holoplenum is to begin at the event horizon of the Planck holosphere. How close to the surface of the Planck holosphere does spacetime actually begin? The overriding constraint is that it *cannot* begin any closer than one Planck length, because anything less than that has no meaning in space–time. Accordingly, the first shell, or "isosphere," must be located exactly *one Planck length from* the central holosphere (fig. 12). Moving radially outward from the central holosphere, each subsequent isosphere enclosing the previous isosphere. They can be visualized as nested, like hollow Russian dolls, or spherical tree rings. Within space, each isosphere is separated from the previous by a distance of 10^{-35} meters, the Planck length.

But of what do these isospheres consist? This isospheric shell cannot itself attain any appreciable depth or spatial thickness due to the Planck length constraint; they cannot be within the explicate order of spacetime while their thickness is below the Planck length. Yet if they are thinner than the Planck length, they are too thin to be in space.

It becomes clear that these isospheric shells are *event horizons* between the implicate order (spacetime) and *the implicate order* (the implicate order). Indicated in the figure, between each shell is a sort of dielectric, not space, but the implicate order. but a quantum vacuum which may also be the locus of Bohm's (1980) "infinite zero-point fluctuations of the 'vacuum field'" (p. 70).

A topology of the implicate order is here coming into focus. A nonlocal, timeless, transcendent domain is to be found at the center, everywhere, in the form of Planck diameter holospheres making up an implicate holoplenum, and enclosing each and every holosphere, nested at quantum multiples from the Planck length to the diameter of the universe, are isospheric shells of the implicate order.

While distributed geometrically within the explicate order, these isospheric quantum shells are not in space, but share in the nonlocality of the implicate order. They provide both a fundamental structure and information backbone bridging the implicate order and the explicate orders. And each isosphere provides a structural capacity for information storage limited only by the Bekenstein bound (discussed in the next section).

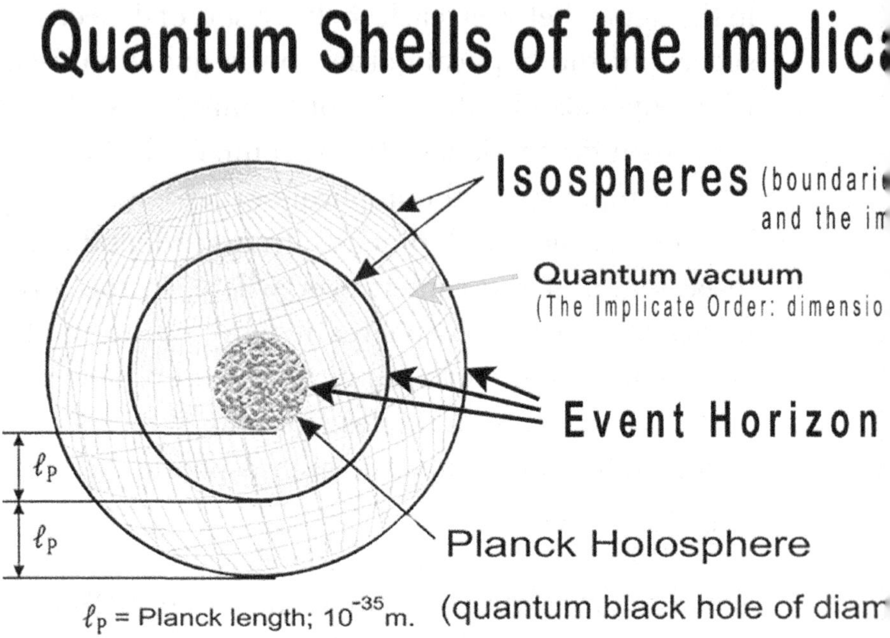

Fig. 12. Quantum Shells of the Implicate Order

As in all black hole phenomena, these isospheres (as well as the central Planck holosphere) are assumed to have an angular momentum (spin) due to turbulence at the event horizon, a highly energetic region which, the holoflux theory posits, are modulated with qubits of information. These spinning, implicate order isospheres project spacetime "objects" (multidimensional holograms) from frequency-phase information within the implicate order.

Spectral frequencies within the implicate order (the frequency domain of communication engineering) can be correlated with sets of isospheres having explicate order diameters corresponding to wavelengths of the same frequency. Thus a complex patterned frequency spectrum within the implicate order can be mirrored in the spatial dimension of the explicate order through the activation of selected ranges of shell isospheres of the implicate.

Having established the mechanism of communication linking Bohm's (1980) implicate and explicate orders, it is now possible to turn to memory storage, that elusive problem which first led Pribram (1971) to his holonomic theory of a brain/mind interface.

A Holoplenum of Quantum Black Holes

As described in the previous section, a holonomic process underlies the Bohmian "Whole" and operates through means of a cosmic plenum of quantum black holes. These quantum black holes present as geometric shells or spheres in space, each one being a boundary between space (outside of the sphere) and the hidden dimensions of the implicate order within the shell boundary (though "within" is a spatial metaphor). Each of these quantum black holes is exactly one quantum in diameter, that is, one Planck length. Thus these quantum-diameter shells have been termed "Planck diameter holospheres."

In geometric terms, these "close-packed spheres" refers to the most tightly packed and highly efficient arrangement of spheres of equal diameter, each of which touches multiple similar adjacent spheres. The entire geometric plenum can be visualized at the very bottom of space, existing everywhere in our spacetime universe, and considered metaphorically as a three-dimensional "ocean floor" of space.

How Holopixels Project the Universe

These inconceivably tiny holospheres act as pixels, or *holopixels*, continually projecting outward into our four dimensional universe of spacetime from the hidden dimensions of the metaverse. Much in the same way that the two-dimensional array of pixels on the back of a plasma display screen projects a streaming video image that we can see, this plenum of quantum holopixels projects the universe into our space and time dimensions. Yet, the region within the boundaries of these black holes contains the multiple hidden dimensions beyond spacetime .

This three-dimensional substrate or plenum exists everywhere and continually projects our spacetime universe that we call the cosmos (fig. 13).

Fig. 13. A Seer at the Intersection of Heaven and Spacetime[7]

In totality, this plenum of holopixels acts much as the two-dimensional plenum of pixels that are embedded in the substrate of a modern large screen plasma video display. Surrounding and radiating outward from each central Planck holosphere throughout space are spherical shells of quantum potential, infinitely thin information-encoded shells of energy, each separated from an inner shell by the quantum exclusion radial distance of one Planck length. These isospheric shells of Bohmian quantum potential extend out to the diameter of the universe itself.

Thus can be visualized topologically an almost infinite series of nested isospheric shells, spread out, at discrete quantum radii from their respective central

Planck holospheres, each bounding Bohm's nondual implicate order.

The global panoramic intersection of these holospheric shells manifests as the projection of a the holonomic universe into spacetime . The cumulative effect of this projection, as regarded by human physicists observing from significantly higher scalar dimensions, is described as "matter." The phenomenon can be understood as a *process of projected creation*, an omnipresent, on-going holographic extrusion of information *from* the implicate order *into* the explicate order, where the various structures of the cosmos (galactic clusters, stars, etc.), the complex unfoldings *into* spacetime , are perceived by the human eye and mind to be three-dimensional, when they are actually holographic projections *from* the implicate order, from the center outward.

Another way of visualizing the projected illusion of spacetime reality from the holoplenum is by expanding upon the metaphor of a flat-panel plasma display (such as the one you may be viewing as you read this).

Consider the human visual threshold for detecting separate images, which lies somewhere between 10 to 12 images per second; the industry standard in the motion picture industry is 24 frames per second (Dorf, 1997, p. 1538). This standard ensures that the presentation of a sequence of projected images will appear to a human viewer as a smooth and continuous motion.

By contrast, if the entire universe flashes in and out of existence at a clock-cycle rate limited only by the Planck time constant of 5.3×10^{-44} seconds, equivalent to a "frame" rate of almost 10^{44} "frames per second," the cosmos would *appear* to be smooth and continuous in all respects even to an electron, and certainly to any human observer of the cosmos, even at quantum dimensions.

The approximate image resolution of a "holoplenum display" obtained by dividing one inch by the Planck length, yields a maximum resolution of 1.584 x 10^{34} holopixels per inch. At such hyper-fine resolution, even a Higgs boson in the 10^{-17} meter dimensional range would appear to be moving smoothly through space.

Holonomic Storage: The Bekenstein Bound

In *Wholeness and the Implicate Order*, Bohm (1980) articulates and develops a "quantum potential" (p. 77) function that projects the explicate spacetime universe out from within an enfolded sub-quantum implicate order. Bohm's quantum potential function is congruent with de Broglie's "Pilot wave" theory of 1927, as both are based upon a conviction that there exist "hidden variables" in sub-quantum regions not accessible to observational exploration using current material science technology (and far beyond the capabilities of the CERN Large Hadron Collider).

The de Broglie pilot wave theory and Bohm's (1980) quantum potential are mathematical attempts to map sub-quantum effects issuing from an implicate order in a domain of "hidden variables" far below the observational capabilities of contemporary material science.

Both theories posit a cybernetic processing of information, simultaneously being cycled from the spacetime world and enfolded into the nondual frequency domain where the accumulating information is processed nonlocally within the implicate order. Driven then by the implicate order, a pilot wave of quantum potential nudges the configurations in spacetime into an altered, slightly new configuration, much as a small tugboat might influence an enormous freighter. If the cosmos operates at its maximum possible clock cycle, as discussed previously, this pilot wave might be seen to operate at the extreme clock cycle rate of the Planck time constant, or 10^{44} Hz.

Somewhere, however, such a cosmic process would require a memory storage repository in spacetime . Regarded as a cybernetic process, the sequence of information feedback and action can be metaphorically imaged in the alchemical Ouroboros, the classical symbol for consciousness, depicted as a snake in a circular configuration eating (or chasing) its own tail. This process can be viewed as the cyclic transfer of information coming in (from the tail) and the resulting action (by the head). A cybernetic feedback loop thus needs data, information, as input. Where then might data be accumulated and retrieved in spacetime at these most fundamental sub-quantum levels, in a Bohmian holonomic universe consisting topologically of the distributed plenum of Planck holospheres, each surrounded by a series of nested quantum isospheres?

One possibility is to consider the information storage potential of an isosphere encoded with granular "bits" of data. In 1970, Jacob D. Bekenstein, then a graduate student working under John Archibald Wheeler (who himself coined the term "black hole"), proposed a novel idea. Bekenstein proposed that there must be an absolute maximum amount of information that can be stored in a finite region of space, and that the Planck constants in quantum theory can be used to determine this limit (Wheeler, 1990). Twenty years later, Bekenstein's theory was extended into what is called the *holographic principle* by Leonard Susskind (1995), which describes how information within any volume of space can be encoded on a boundary of the region. A description of this configuration is presented here by Wheeler (1990) himself, as first related to him by Bekenstein:

> One unit of entropy (information), one unit of randomness, one unit of disorder, Bekenstein explained to me, must be associated with a bit of area of this order of magnitude (a Planck length square). . . . Thus one unit of entropy is associated with each 1.04 x 10^{-69} square meters of the horizon of a black hole. (p. 222)

This proposed upper limit to the information that can be contained upon the surface of a specific, finite volume of space has come to be known as the "Bekenstein bound" (Bekenstein, 1973). Symbolically depicted in Fig. 14 is a topological depiction of the arrangement of information bits, or "qubits," stored on the bounding surface of a spherical volume or isosphere.

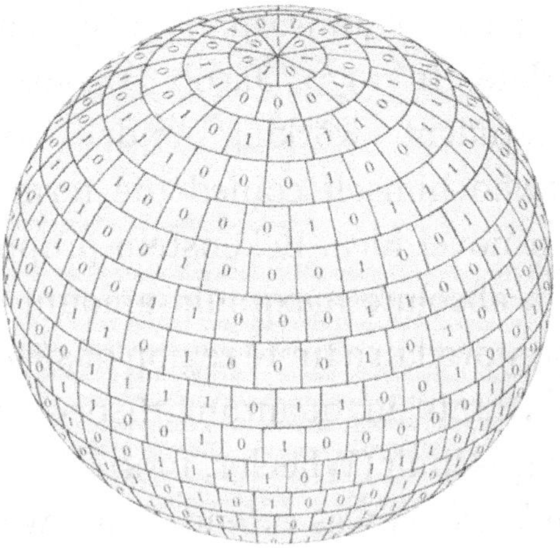

Fig. 14. Planck length qubits on the surface of an isosphere.[8]

This same topological approach to data storage can be applied to human physiology. Using a well-known biological structure as an example, it is possible to calculate the maximum memory storage capacity of a isosphere the size of a single erythrocyte, the ubiquitous red blood cell found throughout the human body. Using Wheeler's approach to determine the number of Bekensteinian equivalent data bits (qubits) on the surface of an isosphere, and using the average diameter of a typical human erythrocyte of 8.1 microns (or 8.1×10^{-5} m), the maximum possible storage capacity on the surface of a single red blood cell can be calculated (Romanes, 1964, p. 137). To obtain this limiting number of bits, the surface area on a spherical shell 8.1 microns in diameter must be divided by 1.04×10^{-69} square meters (which is the Bekenstein unit of entropy, or approximately the square of the Planck length of 1.616199×10^{-35}). The surface area of this erythrocyte-bisected sphere according to this calculation is $4\pi r^2$or $4\pi(8.1 \times 10^{-5})^2 = 4\pi(6.561 \times 10^{-9}) = 8.24 \times 10^{-9}$ square meters. Dividing this by the qubit area of 1.04×10^{-69} square meters yields an estimated maximum storage capacity of 8×10^{60} qubits of storage space for potential information encoding. This is an extremely large data storage capacity, considerably larger than, by contrast, the entire projected capability of the National Security Agency's Utah Data Center, which has been designed, when completed, to have a maximum data storage capacity of twelve exabytes or 12 x 10^{18} bytes (Hill, 2013, para. 7)

Unfolding an Implicate into Spacetime

The ontological understanding of quantum physics that Bohm sought emerges in this model of a sub-quantum, omni-present holoplenum of Planck holospheres, each enclosing a contiguous transcendent region of non-spatial, non-temporal dimensions termed by Bohm “the implicate order” (Joye, 2016).

What are the implications of this model for human consciousness, cognitively operational at temporal and spatial scales vastly larger than those found at these sub-quantum Planck boundaries? To answer this we must first complete the Sub-Quantum cosmological topology of consciousness, and to do this the concept of isospheres, shells within shells, must be considered. Moving outwardly, radially, from the interior bounding event horizon at each central Planck holosphere, can be identified isospheric shells of the implicate order, extruding into space at exact Planck length (quantum) intervals.

This series of concentric shells, each one separated from the next by one Planck length, are isospherical loci of the implicate order (see fig. 5); they extrude into space and they intersect in space with other shells bounding other Planck holospheres in the spacetime holoplenum. It is the cumulative interference effect of the intersection of individual isospheric shells which project images at higher scales, holographically, into three dimensional space.

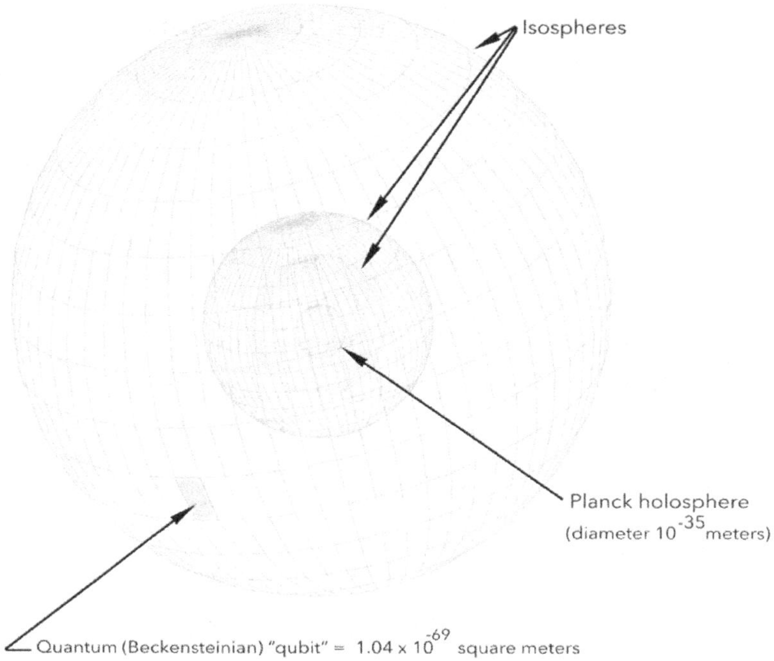

Fig. 15. Topology model of isospheres surrounding a Planck holosphere. Author's figure.

Each isospheric shell, as Bekenstein (1973) determined, has a potentially enormous storage capacity in "qubits" of information encoded on the event-horizon bounding each shell, depending only upon the radius of the shell within the range of 10^{-35} m to 10^{27} m. Accessible simultaneously in both the implicate order and the explicate order, such encoded information provides the data to guide evolving forms as they project into the explicate via the pilot wave mathematics of Bohm's (1980) "many-dimensioned quantum potential" (p. 80). As part of this process, in-formation becomes ex-formation as the implicate order unfurls into the explicate. The plasmoidal forms appearing in spacetime as electromagnetic flux energy are mirrored by and resonate within the implicate order as the dark energy of frequency-phase holoflux.

Within this cosmic geometry of the Sub-Quantum hypothesis can be identified a framework for omnipresent two-way portals, potential bridges between the explicate and the implicate. Here, the deep consciousness of the universe flows in a cyclic, cybernetic, perhaps fractal movement in time and space, moving through an endless, bi-directional process, consciousness involving itself in a dance of transformation.

How then does this topology support human consciousness, thought, and perception in spacetime ? How can electromagnetic frequency plasma in spacetime resonate with holoflux plasma in the implicate order? First of all, the energies must be within the same frequency range in order for maximum interactive resonance to occur. Where the frequencies overlap as they superimpose and interpenetrate one other, resonance occurs.

Resonance is a naturally occurring phenomena characteristic of physical objects or plasma fields extended in spacetime . The resonance effect is seen when objects or complex signal systems exhibit remarkable frequency sensitivity to particular external frequencies, flowing in, through, and around the system, frequencies which approach the "natural resonant frequency" of the object or signal system; perfect resonance occurs where the input frequency and the natural frequency are identical (Feynman, Leighton, &

Sands, Leighton, & Sands, 1964, p. 78). This principle of resonance governs all cybernetic feedback loops, and is a key factor in the design of antennas for electromagnetic transmitting and receiving systems. The goal of antenna design is to construct an antenna that is maximally resonant within a specific narrow frequency range of incoming (external or internal) electromagnetic radiation; when the input frequency and the natural frequency of the antenna coincide or move significantly close to one another, resonance occurs. This simplest and most common antenna design is the dipole antenna, which depends directly on the size (wavelength) of the incoming electromagnetic wave. A dipole antenna is designed to be physically half the size of the incoming wavelength, as stated in a textbook of antenna design:

> A fundamental form of an antenna: length is approximately equal to half the transmitting wavelength. It is the unit from which many more complex forms of antennas are constructed. It is known as a dipole antenna. (Blakeslee, 1972, p. 580)

This same half-wavelength effect governs the design of network communication waveguides applied to fiber-optics in the Internet, where the antenna is the fiber channel itself acting as a waveguide. The waveguide is highly efficient for two reasons: first, as its name implies, the waveguide guides the electromagnetic wave within its channel with maximum efficiency; and secondly, it shields the signal in the channel from external electromagnetic waves. The inner diameter of the hollow waveguide is designed to be equal to exactly half the wavelength of the electromagnetic energy signal shielded by and flowing through the waveguide channel.

Waveguides have been used for over a century both commercially and in research to channel and guide vibrating energy of specific limited frequency ranges; the fiber-optic networks hosting the global internet operate on this principle, channeling electromagnetic radiation at fixed laser frequencies (Dorf, 1997). It was discovered late in the 19th century that circular metallic tubes, or hollow metal ducts, similar to A/C ventilation ducts but much smaller, could be used to channel and guide either sound vibrations in air, or electromagnetic energy in air or vacuum. Without the waveguide, the vibrational energy field is transmitted in all directions, visualized as magnetic lines or arrows emerging from a point at the center of an expanding sphere. This energy disperses outwardly, the magnetic vectored arrowheads pushing into the inside of an infinitely expanding sphere. A waveguide, however, constrains the magnetic component of the wave-front of vibrating energy to one specific linear direction, in parallel with the center of the waveguide, and thus, conceptually, the confined wave itself loses very little power while it propagates along the central axis of the waveguide, like a stream of water emerging from the pinprick of a large, taut, water balloon (Dorf, 1997).

The most common type of fiber-optic cable used in the Internet has a core diameter of 8-10 micrometers and is designed for use in the near infrared (Gowar, 1993, p. 64). The electromagnetic signal wavelength that runs through the global fiberoptic network is powered by highly efficient carbon dioxide lasers, and has a wavelength of 10 micrometers. Coincidentally (or not?), the average human blood capillary diameter is also 10 microns, and blood capillaries are at all times full of carbon dioxide.

Dimensional analysis and a cursory examination of human physiology would immediately suggest two candidates for waveguide systems within the human body: (a) the blood capillary system, and (b) the microtubule system. The corresponding resonant frequency for electromagnetic waves using such waveguides correspond to wavelengths matching the inner diameter of these structures. For blood system capillaries, this corresponds to radiation with a wavelength of 9.3 to 10.0 microns, the average inner diameter of a capillary. For microtubules, the radiation wavelength would be found in a range of 40 nanometers, the inner diameter of the microtubule waveguides. Fig. 16 depicts the location of each of these potential waveguide frequency bands within a wider section of the electromagnetic spectrum.

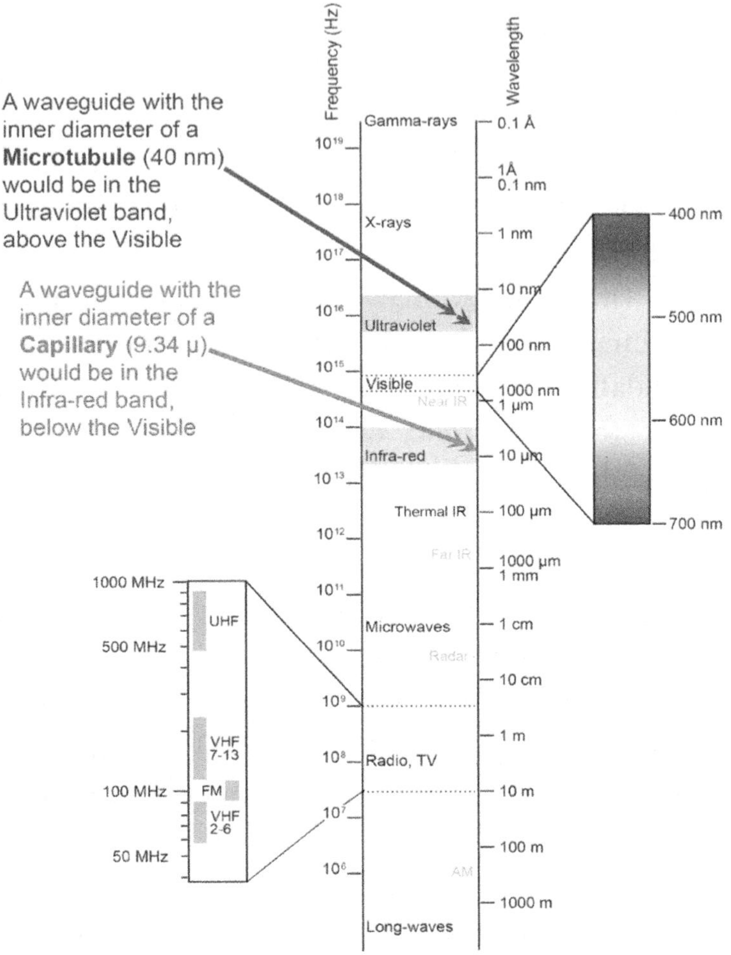

Fig. 16. Microtubules and capillaries as waveguides. Annotations by author; graphic by Jahoe (2012). Reprinted under the terms of a Creative Commons Attribution ShareAlike 2.5 Generic license. Image retrieved from Wikimedia Commons.

There is no impediment to our blood acting as an electromagnetic plasma within the capillary system, and as previously mentioned, the opening page of a textbook on plasma physics reads, "It has often been said that 99%

of the matter in the universe is in the plasma state" (Chen, 2006, p. 1).

In such a model, the entire blood system within the human body can be considered to act as an extensively polarized "super cell" of nonlocal electromagnetic plasma energy, which can then be differentiated from the neuronal brain body of consciousness, itself generated by sequential electrical impulse-driven patterns flowing in the nervous system. Moving charges generate magnetic fields, and ionized human blood flow is no exception: flowing blood plasma results in creation of a magnetic field, and this is in accord with the conjecture of QBD (Dorf, 1997, p. 27).

The circulatory system can be seen as a magnetic plasma composed primarily of ionized red blood cells (erythrocytes) and water molecules, flowing together in complex vortices of blood plasma around every cell and through every capillary of the body (McCraty, 2003). Each erythrocyte is a flexible, annular, bio-concave disk shaped like a doughnut (in geometry, a torus), having a thin webbed center where the hole in a pastry doughnut would be located. The typical outside diameter of a red blood cell is approximately 9 microns, close to the infrared wavelength of 9.6 microns generated by the human body (Turgeon, 2012). The adult human body contains approximately 6 grams of iron, of which 60% is stored throughout the 10^{12} erythrocytes, each of which contains approximately 270 million atoms of ionic iron embedded within transparent hemoglobin in a toroidal locus (Romanes, 1964). Thus each erythrocyte, replete with iron ions embedded in hemoglobin, creates in effect, an ionized iron toroid (Wick, Pinggera, & Lehmann, 2000, p. 6).

Recent studies have also discovered neuronal generation of electromagnetic energy in the near infrared region of the spectrum centered around 10 microns. Radiation emission was repeatedly measured emanating from live crab neurons in extremely narrow, discrete spectral bands within the frequency range corresponding to a spectral region from 10.5 to 6.5 microns (A. Fraser & Frey, 1968).

The implications of this model are considerable: there may exist in nature a unique resonant frequency for each individual human being. It is useful here to step through a topological analysis of the possible functions of a human red blood cell, given its geometry, as a locus of consciousness, and the possible use of the erythrocyte as a locus of memory storage at human biological scales. If, as previously conjectured, the red blood cell has an ideal diameter to resonate electromagnetic radiation in its ferrite-embedded ring at the human infrared wavelength of 10 microns, then it is reasonable to ask if this configuration could accommodate a single unique isospheric frequency (wavelength) for each of the currently 7 billion living humans on the planet. In other words, does this geometry allow for the possibility of each human being to also have a single unique frequency within the infrared electromagnetic radiation that resonates within the human cardiovascular waveguide system? Fig. 7 outlines the topological feasibility of this approach.

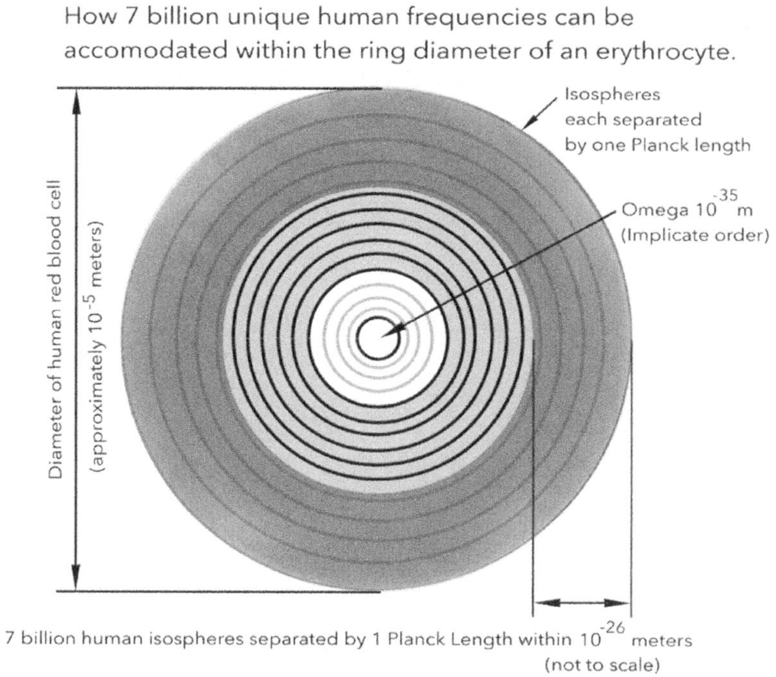

Fig. 17. Isospheric capacity of a single erythrocyte. Author's figure.

Assuming each unique frequency would match its radially unique isosphere, separated by only one Planck length, Figure 17 suggests how 7 billion unique isospheres, each of quantum discrete frequency, might be nested within the geometry of a typical human red blood cell (in the image, a multiple of 7 billion times the Planck length of approximately 10^{-35} m results in an estimated shell thickness of 10^{-26} m). This model supports the feasibility that each living human being might have a unique holospheric frequency, detectable by other human blood cells via the implicate order about which each is centered, and thus provides a possible mechanism for communication, via the mechanisms of resonance, nonlocality, and superposition in the frequency domain of the implicate order.

Sound, Frequency, and the Implicate Order

In contrast with the lack of success others were having in their efforts to establish the mechanism and location of memory engrams, Pribram put forth a theory that the brain mechanics of vision might be seen as holographic processes, projections onto the cortical surface from a frequency domain through the mechanism of frequency superposition and electromagnetic wave interference.[9] Pribram called this "the holonomic brain theory," and postulated the importance of the *frequency domain*:

> Essentially, the theory reads that the brain at one stage of processing performs its analyses in the *frequency domain* . . . a solid body of evidence has accumulated that the auditory, somatosensory, motor, and visual systems of the brain do in fact process, at one or several stages, input from the senses in the *frequency domain*.[10]

Pribram's theory holds that the mathematics of the Fourier transform operates within the brain, modulating electromagnetic fields of flux to create, store, replay, and process holograms within some as yet unidentified physiological context or substrate, but one that Pribram conjectures may be found within the fine-fibered dendritic webs of cerebral cortex regions.[11]

Pribram here takes exception to Bohm's use of the words "flow" or "movement," holding that such words cannot be used to characterize dynamism in a dimension devoid of space or time axes:

> David Bohm (1973) had a concept similar to flux in mind, which he called holomovement. He felt that my use of the term "flux" had connotations for him that he did not want to buy into. I, on the other hand, felt holomovement to be vague in the sense of asking, "what is moving?"[12]

Pribram clearly distinguishes between observations made in the frequency domain and those made in space and time. While the ontological realization of flux in spacetime can be conceptualized in the gyrations of three-dimensional electromagnetic waves, Pribram defines the ontological substrate of spectra in the frequency domain to be pure spectra or holoflux:

> The spectral domain is characteristically a *flux*, composed of oscillations, fluctuations, whereas interference patterns among waves intersect to reinforce or cancel. Holograms are examples of spectra, that is, of a *holoflux*.[13]

This holoflux is essentially the same concept as David Bohm's *holomovement*, though in "Brain and Mathematics,"

Two domains, the time domain (t_d) and the frequency domain (f_d), are of essential importance in the field of information theory and signal communication. On the first page of an undergraduate electrical engineering textbook on network analysis and synthesis, Kuo states:

> In describing signals, we use the two universal languages of electrical engineering—*time* and *frequency*. Strictly speaking, a signal is a function of time. However, the signal can be described equally well in terms of *spectral* or *frequency* information. As between any two languages, such as French and German, translation is needed to render information given in one language comprehensible in the other. Between time and frequency, the translation is effected by the *Fourier series* and the *Fourier integral*.[14]

In his book *Brain and Perception*, Pribram used almost the same words as Kuo in discussing methods by which a Fourier transform operates in mathematical physics. Here Pribram reaches the same conclusion as Kuo:

> It is reasonable to ask: What advantage does the organism gain by processing in the spectral transform domain? The answer is efficiency: The fact that correlations are *so easily achieved* by first convolving signals in the spectral domain and then, inverse transforming them into the spacetime domain. Thus Fast Fourier Transform (FFT) procedures have become the basis of computerized tomography, the CT scans used by hospitals.[15]

The mathematics expressed in the Fourier transform can also be understood as a link between a physics of spacetime (t_d) and an actual frequency domain (f_d). The Fourier and inverse Fourier transforms are shown in fig. 8:

$$f(t) = \int_{-\infty}^{+\infty} X(F$$

Fourier integral transform of a continuous time function into the frequency domain (f_d).

$$f(F) = \int_{-\infty}^{+\infty} x(t$$

Fourier integral transform of a continuous frequency function into the time domain (t_d).

Fig. 18. Fourier Transform/Inverse Transform.[16]

These Fourier transform expressions indicate that any arbitrary function in the timespace domain, $f(t)$, can be transformed into and expressed by an infinite series of frequency spectra functions $X(F)$ in the frequency domain, and conversely, that any arbitrary function in the frequency domain, $f(F)$ can be transformed into and expressed by an infinite series of time functions, $x(t)$.

These Fourier transforms are themselves derived from an underlying series of alternate pure sine and pure cosine waves, as depicted in fig. 9.

$$f(t) = a_0 + \sum_{n=1}^{\infty} \left(a_n \cos \frac{n\pi t}{L} + b_n \sin \frac{n\pi t}{L} \right)$$

Fig. 19. The Fourier Series.[17]

A century after Fourier's death, Norbert Wiener made use of Fourier's transform to model and analyze brain waves, and he was able to detect frequencies, centered within different spatial locations on the cortex, that exhibited auto-correlation. Specific frequencies were found to be attracting one another toward an intermediate frequency, thus exhibiting resonance or "self tuning" within a narrow range of the frequency domain (f_d).[18] This discovery led Wiener to conjecture that the infrared band of electromagnetic flux may be the loci of "self-organizing systems."[19]

> We thus see that a nonlinear interaction causing the attraction of frequency can generate a *self-organizing system*, as it does in the case of the brain waves we have discussed . . . . This possibility of self-organization is by no means limited to the very low frequency of these two phenomena. Consider self-organizing systems at the frequency level, say, of infrared light.[20]

Wiener goes on to discuss such possibilities in biology, where he focuses upon the problems of communication at molecular and primitive cellular levels, specifically on the problem of how substances produce cancer by reproducing themselves to mimic pre-existing normal local cells. Molecules do not simply pass notes to one another, and they do not have eyes, so how do they perceive and how do they communicate? Wiener conjectures:

> The usual explanation given is that one molecule of these substances acts as a template according to which the constituent's smaller molecules lay themselves down and unite into a similar macromolecule. However, an entirely possible way of describing such forces is that the active bearer of the specificity of a molecule may lie *in the frequency pattern of its molecular radiation*, an important part of which may lie *in infrared electromagnetic frequencies* or even lower. It is quite possible that this phenomenon may be regarded as a sort of attractive interaction of frequency.[21]

Perhaps the most fascinating tool for exploring the nature of the interface between the Real–Imaginary domains was developed in 1980 when the Polish–American engineer, Benoit Mandelbrot, created the software to plot an actual image of the two-dimensional interface between spacetime and frequency domains close to the origin (defined as the intersecting point where the Real axis equals zero and the Imaginary axis equals zero). His initial impression, upon seeing the first image, was that the computer program had malfunctioned.[22] Subsequent computer plots assured him that these visual patterns were truly there. Images of this region about the time frequency origin have gained interest worldwide and the region itself has come to be known as the Mandelbrot set (fig. 20). The English mathematical physicist Sir Roger Penrose was so taken by the resulting images that he described them with a sense of almost reverential awe:

> The Mandelbrot set is not an invention of the human mind: it was a discovery. Just like Mount Everest, the Mandelbrot set is just *there*![23]

Fig. 20. Mandelbrot: Origin of real–imaginary axis. Graphic by author.

The Mandelbrot set exhibits remarkable properties. As calculations are done on ever smaller regions on the time–frequency plane, the images appear similar but never completely repeatable, and the viewer begins to sense some sort of biological shapes emerging from this strange world of pure mathematical being. Penrose goes on to say, "The very system of complex numbers has a profound and timeless reality which goes quite beyond the mental constructions of any particular mathematician."[24]

Jung observes the relationship of psyche to matter with the same pattern, noting that they are two different modes of one and the same thing:

> Since psyche and matter are contained in one and the same world, and moreover are in continuous contact with one another and ultimately rest on irrepresentable, transcendental factors, it is not only possible, but also fairly probable, even, that psyche and matter are two different aspects of one and the same thing.[25]

Holoflux, as a total manifestation, is a dual process of reflective consciousness and transcendent awareness. Communication, and consciousness flow between and within the two domains: the implicate and the explicate. Another way of regarding holoflux is to see it as a two-way moving flux: (1) It is an explicate energy (E) moving *inward* into the implicate order at the holospheric event horizon of a quantum black hole, and (2) Moving in the opposite direction, it is an implicate energy (Q) moving *outward* as quantum potential from the event horizon, as holoflux energy transforms from out of the implicate order into the explicate order.

The holoflux theory of consciousness rests upon the assumptions that Bohm's implicate order is real, that the frequency domain is real, and that his active information wavefunction is real, affecting our world of "particles" as it springs from the virtually infinite nonlocal information holoflux within the implicate order.

Like Bohm, electrical engineers separate the world into two domains, which they term the *complex-frequency domain* and the *time domain.* It is our normally perceived spacetime that electrical engineers identify as the *time domain*, but it is in the *frequency domain* that they see the operation of communication signals.[26]

Franklin F. Kuo, an American computer scientist and communications engineer at Bell Laboratories, wrote an electrical engineering textbook, developing the mathematics of communication in *Network Analysis and Synthesis*. In the first chapter of his classic textbook, Kuo describes the two domains, and the mathematical advantage of creating and receiving communication signals within the *frequency* domain as opposed to the *spacetime* domain:

> In the *complex-frequency domain*, the voltage–current relationships for the elements are expressed in *algebraic* equations. Algebraic equations are, in most cases, more easily solved than differential equations. Herein lies the raison *d'être* for describing signals and networks in the *frequency domain* as well as in the *time domain*.[27]

Franklin F. Kuo, an American computer scientist and communications engineer at Bell Laboratories, wrote an electrical engineering textbook, developing the mathematics of communication in *Network Analysis and Synthesis*. In the first chapter of his classic textbook, Kuo describes the two domains, and the mathematical advantage of creating and receiving communication signals within the *frequency* domain as opposed to the *spacetime* domain:

> In the *complex-frequency domain*, the voltage–current relationships for the elements are expressed in *algebraic* equations. Algebraic equations are, in most cases, more easily solved than differential equations. Herein lies the raison *d'être* for describing signals and networks in the *frequency domain* as well as in the *time domain*.[28]

These two orders of electrical engineering theory can be seen to reflect David Bohm's explicate and implicate orders. Bohm's explicate order is associated with the engineering time dimension through which signals are transmitted and received sequentially in time, while Bohm's implicate order can be identified with the electrical engineering *frequency* dimension as a timeless, spaceless domain of pure frequency-phase information.

It has been discovered that these two domains are strongly linked, mathematically, with the operation of the Fourier transform. As Kuo says, "between time and frequency, the translation is effected by the Fourier series."[29] As we have seen in Chapter 4, the Fourier series has also become the bedrock of quantum theoretical calculations, as expressed by David Bohm in the opening of his textbook, *Quantum Theory*:

> It seems impossible to develop quantum concepts extensively without Fourier analysis.[30]

Our thesis, then, is that Kuo's frequency domain, bridged to time by Fourier's mathematics in our twenty-first-century information technology, has an ontological reality, and that there does exist an actual frequency domain, which we also identify with Bohm's hypothesized implicate order.

CONCLUSION: Hidden Dimensions of the Mind

Throughout our lifetimes, each one of us views, interprets, and understands the universe from a truly unique vantage point in space and time. Billions upon billions of moments of experience, thoughts, memories, and sensations accrue in our personal lifetime, resulting in an increasingly unique perspective and the patterns of thought that make each one of us unique. At birth, our small seemingly separate awareness awakens with the merging of two parental streams of consciousness, forming a nexus of awareness that is the new human psyche. This infant awareness opens its inner eye, rising up from the depths of consciousness, becoming a small new island of awareness surrounded by the dark depths of consciousness. Seeded with the inheritance not only of from two parents, but from the experiences of their entire ancestral history, this new, unique awareness begins to accrue experience. New experiences affect the brain as it grows. A new, unique personality, whether labeled purusha, soul or ego, this unique mind continuously changes as it accrues a lifetime of new experiences. How does the human brain structure this new human mind?

One Brain: Two Hemispherical Minds

A clue to the structure of the psyche operating as a human mind within a human brain can be found in observations collected from decades of split-brain research. In the architecture of the psyche, each human mind is a dynamic system generated and sustained by energy flowing through two sides of what is called the cerebrum (fig. 12.1).

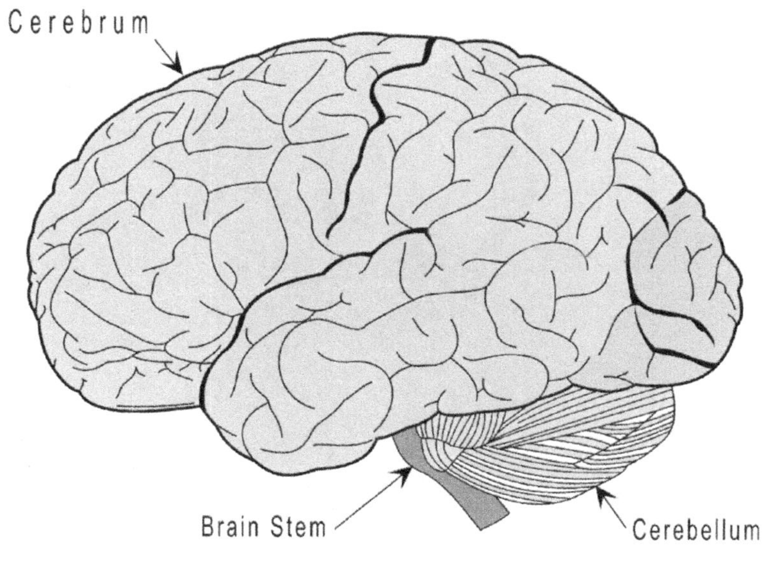

Fig. 12.1. Cerebrum, cerebellum, and brain stem.

Comprising 85 percent of total brain weight, the cerebrum is the largest part of the brain and sits on the top of the Brain Stem. It is also the most recently developed region in the brain's evolutionary history.

The cerebrum is split into two physically separate halves: the left and right cerebral hemispheres (fig. 12.2). The only physical connection between the two is a mass of 300 million axons (nerve fibers) called the *corpus callosum*. After instances in which medical treatment (as in the case of grand mal epileptic seizures) necessitates completely severing this thick white mass of nerve fibers, the frequent post-operative result is that patients have often been observed to exhibit behavior as if they were two entirely distinct (and not always compatible) personalities.

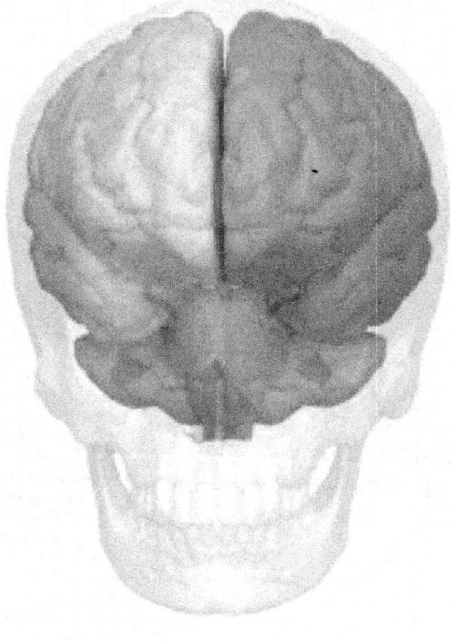

Fig. 12.2. Split brain: view of cerebrum divide.

This phenomenon suggests that what we typically consider our single conscious self is likely made up of two minds that usually, but not always, work together as one, similar to the way two adults function in a marriage. Years of split-brain research suggest that each human being may possess two distinct centers of cognition, or "sub-personalities," which typically function together so seamlessly that they present as a single, unified mind. If a contemporary individual's mind is actually the product of a dynamic relationship between these two sub-personalities, this understanding could have profound implications for the treatment of personality disorders.

A therapist who recognizes that a patient might consist of two differentiated personalities experiencing internal conflict might approach the analysis and treatment of symptoms very differently than if the patient were assumed to have a single, unified personality. This perspective isn't limited to scientific research; it resonates with cultural insights as well. For instance, Rastafarians often use the phrase "I and I" to express the idea that each person embodies two entities—an "I" and another "I"—a duality that modern culture generally overlooks (other than in individuals diagnosed with schizophrenia).

Using the shorthand "LH" and "RH" for "Left Hemisphere" and "Right Hemisphere" personalities, we can observe that in a healthy, well-integrated individual, the LH and RH work cooperatively, sharing common goals and values. However, in individuals with personality disorders, the LH and RH may have conflicting goals and values that require resolution. Problems also arise when one hemisphere dominates the other, suppressing its counterpart.

This understanding complicates the traditional view of two-person, or dyadic, relationships. Previously, it was assumed that such relationships (e.g., between a wife and husband, mother and child, or siblings) involved only two personalities. Consequently, therapists have classically approached these relationships with the goal of addressing the dynamics between two distinct personalities. However, split-brain research suggests that each individual actually brings two distinct personalities into the relationship, leading to a total of four interacting personalities. This means that any one of the four sub-personalities could potentially be in conflict with any of the other three. Thus, the dynamics between two people are far more complex than the formerly assumed interaction between two centers of personality, making it a far more intricate, four-way relationship.

Assume, for example, that a newborn's brain carries holonomic imprints from both parents—perhaps the left hemisphere bears traces of the father's personality, while the right hemisphere carries those of the mother. As the child grows, life experiences are layered onto these initial configurations and integrated through superposition. Thus when two humans interact, it is not just a relatively simple meeting between two individuals, but rather a complex interaction among four mind centers or "hemispherical personalities." The six simultaneous personality relationships are shown in fig.12.3.

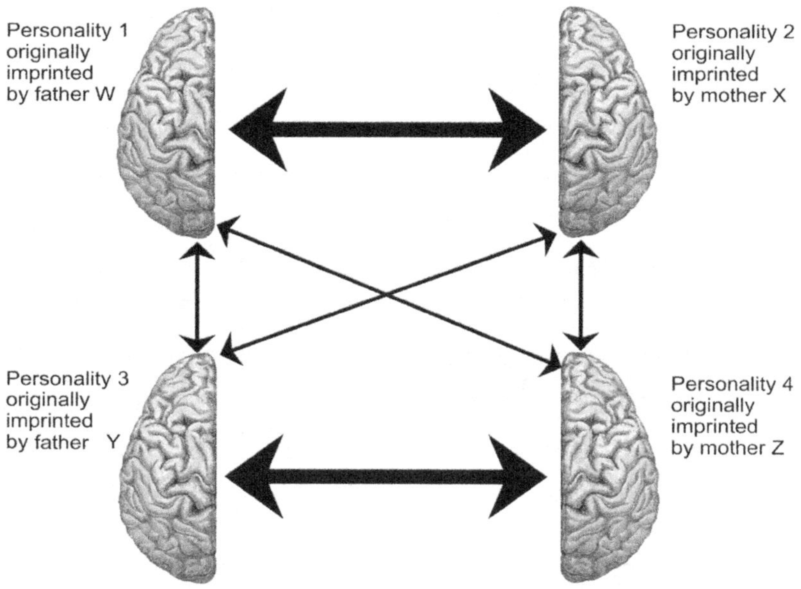

Fig. 12.3. Six personality relationships between two individuals.

While these sub-personalities often align harmoniously, this isn't always the case. Conflicts or incompatibilities between any of the sub-personalities can lead to personality disorders, relationship issues, and challenges within the overall relationship. Wider acceptance of the split-brain multiple personality architecture would have an enormous impact on understanding and treating issues such as schizophrenia, transgenderism, and multiple personality disorder.

In general, the left hemispherical side of the brain has been observed to be responsible for language and speech; because this is how we present ourselves to the world at large, it has been called the "dominant" hemisphere. By contrast, the right hemisphere plays a bigger part in interpreting visual information and spatial processing.

The following table offers some of the observed differences.[31]

Left vs Right Cerebral "Minds"

Left hemisphere: Rational side (linear thinking/male oriented)	**Right hemisphere: Intuitive side** (parallel thinking/female oriented)
Mechanistic; analytically machine-like	Intuitive; perceives patterns
Left hemisphere analyzes (takes apart)	Right hemisphere synthesizes (puts together)
Language, logic, mathematics	Music, humor, mysticism, visual arts
Looks for differences (differentiates)	Looks for similarities (integrates)
Searches for understanding through reasoning	Searches for understanding through intuiting
Proclivity for critical thinking	Proclivity for reading emotions
Fascination with numbers	Fascination with colors, sounds, sensations
Is tightly planned and structured	Is fluid and spontaneous

Parses and analyses information for individual details	Integrates information to obtain a wider picture
Solves problems by logically and sequentially looking at the parts	Problem solves intuitively, searching for patterns and configurations
Communication through talking and writing	Communication through music, drawing, painting, and touch
Tightly restrains expression of emotional feelings	Freely expressess emotional feelings
Memory adapted for recollection of symbols and spoken and written messages	Memory adapted for recollection of color patterns, music, kinetic body positions and sequences
Responds best to audible verbal cues	Responds best to visual cues

Your "mind" can thus be seen as a loosely integrated combination of two separate but superimposed centers of cognitive function and awareness, not always coequal. It is as if two psyches are active simultaneously within a single cranium. These two architecturally separate conscious entities evolve over each individual's lifetime from the initial imprinting passed on from the psyches of mother and father during the moments of procreation. It is thought that the left-hemisphere's psyche holographically inherits the complex of one's father (and his own family predecessors) while the right-hemisphere's psyche is imprinted with the complex from one's mother (and her line of family ancestors).

From that initial imprinting, one's life experiences begin to overlay and supplement the initially inherited configurations of the psyche. As time progresses, one acquires layers of new experience through interactions with the world, and the patterned programming of each hemisphere's actual entity grows even more unique.

It has been widely observed among post-operative individuals who have had the corpus callosum severed, that the two quasi-independent psyches are not always compatible, that is, they do not always seem to agree with one another. In some instances they have been observed to act aggressively at odds with one another.

In most individuals, whether or not the corpus callosum has been severed, one of the two hemispheres usually dominates the other. In a similar way we say a person can be seen to be "right-handed" or "left-handed." One can surmise that the pressures of social conformity in patriarchal societies may further dominance of the left hemisphere's analytical, verbal skills over the right hemisphere's integrative, intuitive strengths.

Modern educational methodologies have arisen to balance the inherent left-right hemisphere attributes through the equal development of both cognitive and creative attributes. The ensuing integration and ensuing harmony is seen to be highly beneficial, resulting in the healthy operation of the composite mind and a healthy psyche. Lack of such integration can lead to dysfunctional operation of the “two minds” within the seemingly single individual.

This suggests the importance of a balanced education early in life, one that offers language skills, mathematical abilities, reasoning and analytical skills to develop the left-hemispherical personality, while, on the other hand, offering training in the visual arts, music, literature, dancing, and social-emotional interaction skills for the right hemisphere. Rudolph Steiner actually developed such a balanced system that has become the foundation of the Waldorf school system now found in many countries. Unfortunately, modern education now stresses logic, reason, and language at the expense of the arts and social skills, an imbalance that is likely reflected in many of the current problems within societies throughout the world.

The Third Mind (beyond the Brain)

In addition to the evidence from studies of patient behavior after corpus callosum severance, both religious tradition and contemplative introspection offer additional supporting evidence that there is a third center of consciousness or “mind” that can be associated with the human psyche, traditionally called soul in the Christian culture, purusha in Hindu culture, and Tao in Chinese culture.

The renowned mystic G. I. Gurdjieff strongly supported the idea of this third mind. Gurdjieff wrote extensively about human’s having multiple mind centers (many “small I’s”) and characterized humans as “three-brained beings.” In 1927, Gurdjieff completed his 1,248-

page allegorical novel Beelzebub's Tales to His Grandson: All and Everything, that includes a fictional account of the firsthand experience of an interstellar traveler visiting the Earth. Early in the book it is noted that humans on this planet have evolved as "three-brained beings." In relating the history of humankind, Gurdjieff covers the essential elements of his system of conscious development and describes how the vast majority of beings on planet Earth live their entire lives in a state of hypnotic "waking sleep." Nevertheless, he tells us, it is possible to transcend to a higher state of consciousness awareness and achieve full human potential through quieting all of one's "little 'I's" and thus allowing the third center of consciousness to rise up into fuller awareness.

With sufficient training, this third mind begins to actively function in the foreground with active links to the other dimensions of the metaverse. This "third-brain" or transcendental mind, operating beyond space and time, is the actual substrate or foundation, the Self as described by Jung, resonating with the two human spacetime mind centers, the left cerebral hemispherical mind and the right cerebral hemispherical mind. In India, this Self is known as the *purusha*, and thought to be the root of our true separate consciousness, emanating directly from Brahma that which is "unchanging, eternal, and pure," and transcending space and time. Each *purusha* is a single seemingly separate actual entity, a diffraction of the One Whole, the Brahma.

One of the goals of the psychotherapeutic process pioneered by Jung is to assist the individual ego to open up to this larger Self, to enable a healthy interrelationship to be established between the normally seemingly isolated, individual ego of the human individual and this Self that bridges the metaverse of dimensions (and the myriad of psychoidal archetypes within these dimensions) in its entirety. Jung termed such a healthy connection between ego and archetypes an "individuation," a state in which the ego operates in a balanced, healthy, whole manner as an integral part of an ocean of collective psychoids that make up the greater unitary metaverse.

Many contemplative traditions teach individuals how to practice mental exercises that can lead to the development of the capacity to silence normal mental activities in order to open the portal to this wider dimensional metaverse. During the period of sustained contemplative silence of the two cerebral "laptop-minds," a previously unperceived panorama arises into awareness, a scintillating holonomic flow of holoflux that some have called "cosmic consciousness" (Bucke) or "one taste" (Wilber). The new awareness is not merely visual, but a combination of sensations perceived through direct contact with multiple dimensions beyond normal human sensory systems. At some moment during meditation, one experiences a "rupture of plane" (McKenna) as the formerly dualistic cognitive identity fades into the background, replaced with a deep and immediate sense of having joined a cosmic stream of flowing energies.

This flow of holoflux energy is not just a movement in time and space, but is a holonomic flow within and between the many dimensions beyond spacetime, including the deepest source dimensions that are the foundation of all, as discussed here in this dialog between David Bohm and Jiddhu Krishnamurti:

Bohm: Would you say energy is a kind of movement?

Krishnamurti: No, it is energy. The moment it is a movement it goes off into this field of thought.

Bohm. We have to clarify this notion of energy. I have also looked up this word. You see, it is based on the notion of work; energy means, "to work within."

Krishnamurti: Work within, yes.

Bohm: But now you say there is an energy which works, but *no movement*.

Krishnamurti: Yes. I was thinking about this yesterday—not thinking—I realized *the source* is there, uncontaminated, *non-movement*, untouched by thought, it is there. From that these two are born. Why are they born at all?

Bohm: One was necessary for survival.

Krishnamurti: In survival this—in its totality, in its wholeness—has been denied, or put aside. What I am trying to get at is this, Sir. I want to find

> out, as a human being living in this world with all the chaos and suffering, *can the human mind touch that source in which the two divisions don't exist?*—and because it has touched this source, which has no divisions, it can operate without the sense of division.[32]

Thus we are not the "person" we think we are; in fact, we are much, much more. We are the universe looking out from a particular perspective into time and space, observing and interacting with itself.

Opening up to the Self and the Metaverse

So how does one begin to open up one's own portal of consciousness to the wider Self and thus to the metaverse? There are many techniques that have been discovered down through the ages in most, if not all cultures, passed down as skillful means in religious and shamanic traditions. Some traditions have identified the ingestion of plant entheogens as a beneficial approach for opening one's consciousness to the initial experience. Other methods stress the efficacy of physical exercises such as yoga, tai chi chuan, and even dervish dancing for catalyzing entry into the silence. The primary emphasis in all of these approaches is the focus upon learning to enter a state of cognitive silence. Only through the silencing of the normal mind (called "the monkey mind" in some traditions) can the experience of the deeper pervasive silence arise.

To explore the normally hidden dimensions of consciousness, one must, through practice, learn to quiet the normal activity of the brain, sometimes referred to in yoga as "quieting the monkey mind." This requires "letting go" of one's normal cognitive flow (i.e., random thoughts, internal verbal dialog, memory sequences) while simultaneously reducing sensory awareness of warmth, cold, discomfort, and external sounds. Texts of classical yoga recommend meditating in silent caves or the desert, while my own practice for many years has been to meditate in a dark room at night while wearing earplugs. My mentor John Lilly practiced meditation for long periods floating in body-temperature salt water in a lightproof, soundproof chamber of his own design, often under the influence of ketamine or LSD.

Eventually, with consistent practice, one can significantly diminish the flow of thoughts and distraction of external sensations. At some point, emerging from the silence, a new mode (Ken Wilber calls it a new "flavor") of consciousness begins to enter awareness, an unusual but unmistakable sense of joining a living network of consciousness, perhaps even more than one distinct network. There is a pronounced sensation of actually touching something and being touched by something.

At a certain moment, one's field of awareness opens up to something that is much bigger, wider, and deeper than normal sensory input, and though first one may sense a fear of this unknown vastness, the fear is soon replaced with a sense of belonging, of being welcomed. In some traditions this is termed the experience of the primordial consciousness, in Sanskrit the *dharmadhātu*, that underlies the entire fabric of reality, the doctrine of the Void, or *śūnyatā*, that is the goal of contemplative practices in Tibetan Dzogchen, Chan, and other Buddhist traditions. This experience of the Void is here described by the psychiatrist Stanislav Grof:

> The Void, primordial Emptiness and Nothingness is consciousness itself. The Void has a paradoxical nature; it is a vacuum, because it is devoid of any concrete forms, but it is also a plenum, since it seems to contain all of creation in a potential form. According to Ervin Laszlo, the Void is a subquantum field which is the source of all creation and in which everything that has ever happened remains holographically recorded. Laszlo equates this field with the concept of quantum vacuum that has emerged from modern physics.[33]

Bohm, from the viewpoint of quantum physics, also clearly stated that the void is not simply an empty vacuum but a plenum: "Space is not empty. It is full, a plenum as opposed to a vacuum, and is the ground for the existence of everything, including ourselves. The universe is not separate from this cosmic sea of energy."[34]

In order to reach this experience of the void, our memories, thoughts, and external sensations need to be ignored and allowed to fade away. My own first experience of opening to these hidden dimensions occurred (described at length in chapter 2) in a quiet inner room of my fifth-floor walkup apartment, while doing a hatha yoga pose in the silent darkness late at night. Suddenly from out of the silence, a loud, pure, high-pitched whistling sound arose into my awareness. Over the ensuing years my perception of this sound has grown richer, with increasingly complex "subtle sounds" and various feelings of awareness and intuitions, inexpressible in terms of space and time and words.

I later discovered that this subtle awareness is not quite as rare as I had thought. Recent medical research has determined that 15 percent of the adult population experiences what is called *tinnitus* at some time in their lives. It is thought to be to be a widespread issue that affects millions of people globally. Modern medicine treats tinnitus as a disease, as it is often disconcerting and anxiety-provoking to those who begin to hear the tones, sometimes quite loud. But what if tinnitus is not a disease? Could it simply be the beginning of an opening to the hidden dimensions of the void? What if people, instead of running from the phenomenon, choose instead to begin exploring, even cultivating these sounds, rather than trying to suppress them in fear after being told that they are somehow a physical disease? Imagine the first sea creatures responding with fear at the unknown new sensory input of the sensation of light as their rudimentary eyes began to open awareness to the sun. My own conjecture, as a computer and electrical engineer, is that this substrate of sound frequencies that I hear when my brain quiets down, is an opening to a sort of cosmic machine language, an inner cosmic network operating at extremely high frequencies (at very small, perhaps sub-quantum dimensions). This might explain my own experiences on ayahuasca of sensing a link or portal opening up to a wider awareness of the universe, one that seemed to examine my body on many levels and somehow then tuned up various systems within me.

Conclusions and Recommendations

In our search for guidance to help us understand and better navigate the profound challenges we face, both as a planet and as individuals, we have turned in this book to the rich maps of consciousness developed by Carl Jung, Teilhard de Chardin, and Gustav Fechner. Each of their cosmologies locates the consciousness of an individual human within a wide spectrum of conscious entities that operate outside of ordinary human awareness in vast, largely uncharted dimensions. Much of their insight and detailed theories were developed through direct personal experiences of consciousness. Their common methodology was to engage regularly in direct contemplative observation of the movement of their own minds, a technique pioneered by Wilhelm Wunt (1832-1920) that he termed *introspection.* Unfortunately the practice of introspection was eventually dismissed and even denigrated by modern scientific materialists who have grown ever more entrenched in their myopic insistence that the *only* data that can be accepted *must be* of objective, repeatable, recordable behavior, and definitely not obtained through "observing the mind with the mind itself." Thus the rich integrated maps developed by Jung, Teilhard, and Fechner, potentially answering many questions about consciousness and pointing the way toward further research, were often disregarded and frequently excluded from "serious" scientific discussion.

But recently the theories of Jung, Teilhard, and Fechner have been gaining new recognition and respect due to advances in particle physics and mathematics. String theory and high-energy particle experiments now support the existence of at least seven dimensions beyond spacetime, and developments in artificial intelligence and neuronal networking highlight the possibility of vast, interconnected networks of conscious agents (psychoids) extending weblike throughout multiple dimensions of reality. The writings of Jung, Teilhard, and Fechner all suggest that the human brain and mind do posses the capacity to tune into and to join these other levels or networks of consciousness. Once a link with other modalities of consciousness has been established, whether through meditation, entheogens, prayer, or technological advancements (e.g., so-called "God helmets," or brain implanted emf devices), the connection, tenuous at first, can be strengthened through regular practice, eventually becoming a channel for communication with other psychic entities for the potential reception of knowledge, guidance, and insight, from other sources of consciousness.

The idea that consciousness may manifest in frequency vibrations that can enable links to hidden conscious entities is not a new one. The Swedish philosopher Emanuel Swedenborg (1688-1772) was a trained scientists who wrote extensively on his regular communication with conscious entities that are beyond normal human perception. In his 1718 dissertation, "On the Mechanism of the Operation of the Soul and the Body," Swedenborg developed a theory of vibrations or subtle movements—what he referred to as "tremulations"—as a way of understanding how thoughts, sensations, and psychic influences could affect the physical body. In his writings, particularly those that blend his scientific knowledge with mystical theology, he often referred to these vibrational phenomena as a medium for the transfer of spiritual energies or divine influence to the material world.

There are accounts from Emanuel Swedenborg's contemporaries, including his servants, that they claimed to have witnessed him conversing with invisible spirits. These stories gained attention during the later part of Swedenborg's life, when he became deeply involved in his internal experiences and began to write extensively on the subject of his interactions with the spiritual world. Swedenborg himself claimed that from around 1744 onward, he was able to converse directly with angels, spirits, and even deceased individuals from the afterlife.

Accordingly, I encourage you to join with the many explorers of consciousness such as Swedenborg, Jung, Fechner, Teilhard de Chardin and many more in establishing connections between humanity and denzins of the hidden dimensions by forging connections that reach out into the metaverse through regular contemplative practice. Learn to tune your mind as an instrument that can become capable of making that initial contact with new dimensions. At a minimum, your ability to relax into a sustained silence, free from spontaneous mental processes, will give your cerebral cortex a rest and quiet the activity of your two cerebral hemispheres, during which your cortex will no longer be acting as a Faraday cage to shield your inner brain regions from external signals. The psyche then has the possibility of opening out into a deeper awareness of new dimensions beyond the isolated self (fig. 12.4.).

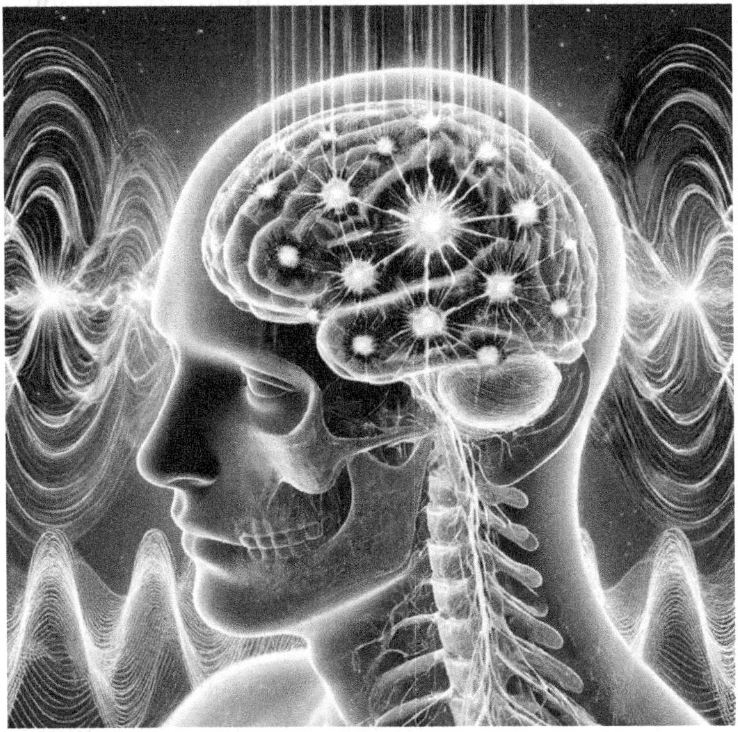

Fig. 12.4. Neurons resonating with hidden dimensions of consciousness.

With effort and intent, one can establish links to rich networks of consciousness (Teilhard's noospheres) that can enhance one's ability to receive intuitive knowledge, while also maintaining a calm, tranquil, centered state of being in the midst of a world filled with anxiety, change, and confusion. Such a regular practice of meditation and contemplation can enrich your being while simultaneously offering one of the few real ways that an individual can open new avenues of possibility to join in solving the many problems that now engulf our civilization on the planet Earth.

I would like to close with a quote from Gustav Theodor Fechner:

> Beside our consciousness there is still more consciousness, that over and above all individual consciousness there is a broader and higher consciousness with broader and higher content, a consciousness which on the side upon which it excels and surpasses our consciousness represents the outward world by which our consciousness is determined, and ties together all individual consciousness by common situations and effective relationships, the highest unity of which is found in the last knot, love.[35]

References

Adelman, George ed. *Encyclopedia of Neuroscience, Vol. 1*. Boston: Birkhäuser Boston, 1987.

Allaby, Michael, and Ailisa Allaby, eds. *A Dictionary of Earth Sciences*. 2nd ed. New York: Oxford University, 1999.

Anderson, Mark. "Do Vibrations Help Us Smell?" *Scientific American*, April 1, 2013. https://www.scientificamerican.com/article/do-vibrations-help-us-smell/

Assmann, Jan. 2008. *Of God and Gods: Egypt, Israel, and the Rise of Monotheism*. Madison: University of Wisconsin Press.

Sri Aurobindo Ghose. *Hymns to the Mystic Fire*. Pondicherry: Sri Aurobindo Ashram, 1972.

Aurobindo, Sri. *The Life Divine*. First published in the monthly review, *Arya* (1914–1920). Pondicherry: Sri Aurobindo Ashram Trust, 1990.

———. *The Upanishads: Texts, Translations and Commentaries*. Pondicherry: Sri Aurobindo Ashram, 1972.

———. *Sri Aurobindo on the Tantra*. 3rd ed. Pondicherry: Dipti, 1972.

———. *The Synthesis of Yoga*. Pondicherry: Sri Aurobindo International University Centre, 1955.

———. *Tales of Prison Life*. Pondicherry: Sri Aurobindo Ashram, 1997.

Azcel, Amir D. *The Jesuit and the Skull: Teilhard de Chardin, Evolution, and the Search for Peking Man*. New York: Riverhead Books, 2007.

Bagchi, P.C. "Evolution of the Tantras." In *The Cultural Heritage of India, Vol. IV,* edited by Haridas Bhattacharyya. Calcutta: The Ramakrishna Mission Institute of Culture, 1937.

Blofeld, John. *Tantric Mysticism of Tibet: A Practical Guide to the Theory, Purpose, and Techniques of Tantric Meditation*. Boston: E.P. Dutton, 1970.

Bloom, Allan. *The Republic of Plato: Translated with Notes and an Interpretive Essay*, 2nd ed. New York: Harper Collins, 1968.

Bohm, David. *Wholeness and the Implicate Order*. London: Routledge, 1980.

Bohm, David, and F. David Peat. 1987. *Science, Order, and Creativity*. London: Routledge.

Bryant, Edwin F. *The Yoga Sutras of Patañjali: A New Edition, Translation, and Commentary with Insights from Traditional Commentators*. New York: North Point Press, 2009.

Bailes, K.E. *Science and Russian Culture in an Age of Revolutions: V.I. Venadsky and His Scientific School, 1863-1945*. Bloomington: Indiana University Press, 1990.

Bailey, Alice. *Light of the Soul: A Paraphrase of the Yoga Sutras of Patañjali*. New York: Lucis Publishing, 1988.

Bailey, Gregory. *The Study of Hinduism*. Columbia: The University of South Carolina Press, 2003.

Barnhart, Bruno. "Christian Self-Understanding in the Light of the East." In *Purity of Heart and Contemplation: A Monastic Dialogue Between Christian and Asian Traditions*, edited by Bruno Barnhart and Joseph Wong, 291–308. New York: Camaldolese Hermits of America, 2001.

Berg, Jerome S. *Broadcasting on the Short Waves, 1945 to Today*. London: McFarland & Company, 2008.

Brown, Raphael. *The Little Flowers of St. Francis*. New York: Hanover House, 1958.

Bryant, E.F. *The Yoga Sutras of Patañjali: A New Edition, Translation, and Commentary with Insights from Traditional Commentators*. New York: North Point Press, 2009.

Chaudhuri, H. *Philosophy of Integralism*. Pondicherry: Sri Aurobindo Pathamandir, 1954.

Chen, Frances F. *Introduction to Plasma Physics and Controlled Fusion: Vol 1. Plasma Physics*. 2nd ed. New York: Springer, 2006.

Cuénot, Claude. *Teilhard de Chardin: A Biographical Study*. London: Burnes & Oates, 1965.

Cunningham, Lawrence S. and Keith J. Egan. *Christian Spirituality: Themes from the Tradition*. Mahwah, New Jersey: Paulist Press, 1996.

de Lubac, Henri. *The Religion of Teilhard de Chardin.* Translated by René Hague. New York: Desclée, 1967.

de Terra, Helmut. *Memories of Teilhard de Chardin.* New York: Harper & Row, 1964.

Düdjom Lingpa. *Heart of the Great Perfection.* Vol. I of *Düdjom Lingpa's Visions of the Great Perfection.* 3 vols. Foreword by Sogyal Rinpoche. Translated by B. Alan Wallace. Somerville, MA: Wisdom Publications, 2015.

Duffy, Kathleen. *Teilhard's Struggles: Embracing the Work of Evolution.* Maryknoll, NY: Orbis Books.

Dyczkowski, Mark S.G. *A Journey in the World of the Tantras.* India: Indica Books, 2004.

Dzogchen Ponlop Rinpoche, *Mind Beyond Death.* Ithaca, NY: Snow Lion, 2008.

Dzongsar Jamyang Khyentse Rinpoche. "The Wheel of Life." *Gentle Voice: A Newsletter of Siddhartha's Intent,* April 2005. http://www.siddharthasintent.org/assets/Global-Files/GentleVoice/GV23.pdf.

Eliade, Mircea. *Yoga: Immortality and Freedom.* Princeton: Princton University, 1954.

Eliot, Thomas Stearns. 1943. *Four Quartets.* New York: Harcourt Brace.

Fechner, Gustav. *Elements of Psychophysics* ~~(1860)~~. Translated by Helmut E. Adler. New York: Holt,

Rinehart, and Winston, 1966. First published in 1860.

Fechner, Gustav Theodor. *Religion of a Scientist: Selections from Gustav Theodor Fechner*. Edited by W. Lowrie. New York: Pantheon, 1946.

Feurstein, Georg. *The Philosophy of Classical Yoga*. Manchester: Manchester University Press, 1980.

Feynman, Richard, Robert Leighton, and Matthew Sands. *The Feynman Lectures on Physics. Volume 1*. Massachusetts: Addison-Wesley, 1964.

Flood, Gavin. *The Tantric Body: The Secret Tradition of Hindu Religions*. London: I.B. Tauris, 2006.

Fox, Matthew. *The Coming of the Cosmic Christ*. New York: Harper One, 1988.

Franz, M.-L. "The Process of Individuation." In *Man and His Symbols*, edited by C. G. Jung and M.-L. Franz, 158–229. New York: Random House, 1964.

Garrison, Omar. *Tantra: The Yoga of Sex*. Hong Kong:Causeway Books, 1964.

Gokhale, Pradeep P. "Interplay of *Sāṅkhya* and Buddhist Ideas in the Yoga of Patañjali," in *Journal of Buddhist Studies*, (Sri Lanka and Hong Kong), Vol. XII, 2014-15, 107-122

Gonda, J. *History of Ancient Indian Religion, Vol. 4: Selected Studies*. Leiden: E.J. Brill, 1975.

Griffiths, Bede. *A New Vision of Reality: Western Science, Eastern Mysticism, and Christian Faith*. Springfield, IL: Templegate Publishing.

Griffiths, Bede. *Vedanta and Christian Faith.* 2nd ed. Middletown, CA: Dawn Horse Press, 1991.

Gurdjieff, G. I. *Beelzebub's Tales to His Grandson or An Objectively Impartial Criticism of the Life of Man.* New York: Harcourt, 1950.

Haisch, Bernard. *The Purpose-Guided Universe: Believing in Einstein, Darwin, and God.* New Jersey: Career, 2010.

Hariharānanda, P. *Kriya Yoga: The Scientific Process of Soul Culture and the Essence of All Religion.* Delhi: Motilal Banarsidass, 2006.

Hariyappa, H. L. 1953. *Rigvedic Legends Through the Ages.* Bombay: Mysore University.

HeartMath Institute. "Energetic Communication." Chapter 6 of Science of the Heart: Exploring the Role of the Heart in Human Performance. HeartMath Institute. Accessed October 19, 2020 https://www.heartmath.org/research/science-of-the-heart/energetic-communication/.

Heehs, Peter. *The Lives of Sri Aurobindo.* New York: Columbia University Press, 2008.

Heile, Frank. "Time, Nonduality and Symbolic versus Primary Consciousness." Lecture presented at the Science and Nonduality Conference, San Rafael, CA, October 2011.

Hoffman, *The Case Against Reality: Why Evolution Hid the Truth from Our Eyes.* New York: W.W. Norton, 2019.

Huxley, Aldous. *The Doors of Perception.* New York: Harper & Brothers, 1954.

Jaynes, Julian. *The Origin of Consciousness in the Breakdown of the Bicameral Mind.* Boston: Houghton Mifflin, 1990. First published 1976.

Jibu, Mari, and Kunio Yasue. *Quantum Brain Dynamics and Consciousness.* Philadelphia: John Benjamins, 1995.

Jinarajadasa, C. *The Hidden Work of Nature.* Kessinger Publishing, 2010.

Jones, Roger S. *Physics for the Rest of Us: Ten Basic Ideas of Twentieth-Century Physics that Everyone Should Know . . . and How They Have Shaped Our Culture and Consciousness.* New York: Barnes and Noble, 1999.

Joye, Michael R. "The Philosophy, Practice, and History of Tantra in India." Master's Thesis, California Institute of Asian Studies, 1978. ProQuest (1311340).

Joye, Shelli R. *Developing Supersensible Perception: Knowledge of the Higher Worlds through Entheogens, Prayer, and Nondual Awareness.* Rochester, VT: Inner Traditions, 2019.

———. *The Electromagnetic Brain: EM Theories on the Nature of Consciousness.* Rochester, VT: Inner Traditions, 2020.

———. *The Little Book of Consciousness: Holonomic Brain Theory and the Implicate Order*. Viola, CA: The Viola Institute, 2017.

———. *The Little Book of the Holy Trinity: A New Approach to Christianity, Indian Philosophy, and Quantum Physics*. Viola, CA: The Viola Institute, 2017.

———. "The Pribram-Bohm Holoflux Theory of Consciousness: An Integral Interpretation of the theories of Karl Pribram, David Bohm, and Pierre Teilhard de Chardin." PhD diss., California Institute of Integral Studies, 2016. ProQuest (10117892).

———. "The Pribram–Bohm Hypothesis." *Consciousness: Ideas and Research for the Twenty-First Century* 3, no. 3 (2016): Article 1. Accessed October 18, 2020, https://digitalcommons.ciis.edu/conscjournal/vol3/iss3/1.

———. *Sri Aurobindo: Quantum Physics and Consciousness*. Viola, CA: The Viola Institute, 2018.

———. *Sub-Quantum Consciousness: A Geometry of Consciousness Based Upon the Work of Karl Pribram, David Bohm, and Pierre Teilhard De Chardin*. Viola, CA: The Viola Institute, 2019.

———. *Teilhard's Hyperphysics: Energy and the Noosphere*. Viola, CA: The Viola Institute, 2020.

———. *Tuning the Mind: The Geometries of Consciousness*. Viola, CA: The Viola Institute, 2017.

Jung, C.G. "The Structure and Dynamics of the Psyche," in *The Collected Works of C.G. Jung*, vol. 8. Translated by R.F.C. Hull. Princeton: Princeton University Press, 1960.

Jung, C.G., and M.-L. Franz, eds. *Man and His Symbols*. New York: Random House, 1964.

Kavanaugh, K., trans. *Collected Works of St. John of the Cross*. Washington, DC: Institute of Carmelite Studies, 1991.

Khanna, Madhu. *Yantra: The Tantric Symbol of Cosmic Unity*. London: Thames & Hudson, 1979.

King, Ursula. *Pierre Teilhard de Chardin: Writings Selected with an Introduction by Ursula King*. New York: Orbis Books, 1999.

———. *Spirit of Fire: The Life and Vision of Teilhard de Chardin*. New York: Orbis Books, 1996.

Kyczkowski, Mark S.G. *The Doctrine of Vibration: An Analysis of the Doctrines and Practices of Kashmir Shaivism*. Albany: State University of New York Press, 1987.

Kuo, F. *Network Analysis and Synthesis*. New Jersey: Bell Telephone Labs, Inc., 1962.

Le Cocq, Rhoda P. *The Radical Thinkers: Heidegger and Sri Aurobindo*. Pondicherry: Sri Aurobindo Ashram Press, 1969.

Leary, Timothy, and Ralph Metzner, and Richard Alpert. *The Psychedelic Experience: A Manual Based Upon the Tibetan Book of the Dead*. New York: University Books, 1964.

———. *Flashbacks: A Personal and Cultural History of an Era*. New York: Tarcher, 1983.

LeLoup, Jean-Yves. *Being Still: Reflections on an Ancient Mystical Tradition*. Translated and edited by M. S. Laird. Mahwah, NJ: Paulist Press, 2003.

Leroy, Pierre. "Teilhard de Chardin: The Man." Introduction to *The Divine Milieu,* by Teilhard de Chardin, 13–42. New York: Harper & Row, 1960.

Lindorff, D. *Pauli and Jung: The Meeting of Two Great* Minds. Illinois: Quest Books, 2004.

Lilly, John. *The Deep Self: Consciousness Exploration in the Isolation Tank*. New York: Simon & Schuster, 1977.

———. *The Mind of the Dolphin: A Nonhuman Intelligence*. New York: Doubleday, 1967.

———. *Programming and Metaprogramming in the Human Biocomputer: Theory and Experiments*. New York: The Julian Press, 1972.

Lu K'uan Yu. *Taoist Yoga: Alchemy & Immortality*. Boston: Weiser Books, 1973.

Malinski, Tadeusz. *Chemistry of the Heart*. Athens: Ohio University Biochemistry Research Laboratory, 1960.

McCraty, Rollin, Annette Deyhle, and Doc Childre. 2012. "The Global Coherence Initiative: Creating a Coherent Planetary Standing Wave." *Global Advances in Health and Medicine* 1 (1): 64–77.

McFadden, Johnjoe. "The Conscious Electromagnetic Information (CEMI) Field Theory: The Hard Problem Made Easy." In Journal of Consciousness Studies 9, no. 8 (2002): 45-60.

McFadden, Johnjoe. "Synchronous Firing and Its Influence on the Brain's Electromagnetic Field: Evidence for an Electromagnetic Field Theory of Consciousness," *Journal of Consciousness Studies.* vol. 9, no. 4 (2002): 23.

McKenna, Terence. *True Hallucinations: Being an Account of the Author's Extraordinary Adventures in the Devil's Paradise.* New York: HarperCollins, 1993.

Merrell-Wolff, Franklin. *The Philosophy of Consciousness Without an Object.* New York: Julian Press, 1973.

Morgan, Conway Lloyd. *Emergent Evolution: Gifford Lectures, 1921–22.* New York: Simon & Schuster, 1978.

Mukharji, P.B. "Introduction." In Swami Pratyagatmananda Saraswati, *Japasutram: The Science of Creative Sound.* Madras: Ganesh & Co., 1971.

Nagel, Thomas. 1974. "What is it like to be a bat?" *Philosophical Review* 83:435–451.

Netter, F.H. *The CIBA Collection of Medical Illustrations, Vol. I: The Nervous System*. Summit, NJ: CIBA, 1972.

Panikkar, Raimon. *The Rhythm of Being: The Gifford Lectures*. New York: Orbis Books, 2010.

Pandit, M.P. *Lights on the Tantra*. Madras: Ganesh & Co, 1957.

———. *Studies in the Tantras and the Veda*. Madras: Ganesh & Company, 1973.

Parker, R.C. "The Use of Entheogens in the Vajrayāna Tradition: A Brief Summary of Preliminary Findings Together with a Partial Bibliography," retrieved from Vajrayāna.faithweb.com, 2007.

Pearce, J.C. *The Biology of Transcendence*. Rochester, VT: Park Street Press, 2002.

Penrose, Sir Roger. *The Emperor's New Mind: Concerning Computers, Minds and the Laws of Physics*. Oxford: Oxford University Press, 1989.

Persinger, Michael. *Spacetime Transients and Unusual Events*. Chicago: Nelson-Hall, 1977.

Persinger, M., and J.N. Booth. "Discrete Shifts Within the Theta Band Between the Frontal and Parietal Regions of the Right Hemisphere and the Experience of a Sensed Presence." Journal of Neuropsychiatry 21, no. 3 (2005): 279–83. doi:10.1176/appi.neuropsych.21.3.279.

Pine, R. *The Heart Sutra*. Berkeley, CA: Counterpoint, 2004.

Pockett, Susan. *The Nature of Consciousness: A Hypothesis*. Nebraska: Writers Press, 2000.

Powell, A. E. *The Etheric Double*. Madras: The Theosophical Publishing House, 1925.

Rele, Vasant. *The Mysterious Kundalini: The Physical Basis of the 'Kundalini Yoga'*. Bombay: D.B. Taraporevala Sons & Co., 1927.

Reninger, Elizabeth. "The Amazing Pineal Gland." About.com, March 27, 2008, http://taoism.about.com/b/2008/03/27/the-amazing-pineal-gland.htm. Webpage defunct.

Romanes, G.J. ed. *Cunningham's Textbook of Anatomy, 10th Ed.* London: Oxford Press, 1964.

Roy, Dilip Kumar. *Sri Aurobindo Came to Me*. Pondicherry: Sri Aurobindo Ashram Trust, 1952.

Ruhenstroth-Bauer, G. "Influence of the Earth's Magnetic Field on Resting and Activated EEG Mapping in Normal Subjects." *International Journal of Neuroscience*, vol. 73, no. 3-4 (June 1993): 331–49.

Samson, Paul R., and David Pitt, eds. 1999. *The Biosphere and Noosphere Reader: Global Environment, Society and Change*. New York: Routledge.

Shannon, C.E. "A Mathematical Theory of Communication." *Bell System Technical Journal*, no. 27 (July 1948): 623-656.

Sheldrake, Rupert. *Morphic Resonance: The Hypothesis of Formative Causation*. London: Blond & Briggs, 1981.

Sher, Leo. "Neuroimaging, Auditory Hallucinations, and the Bicameral Mind." *Journal of Psychiatry & Neuroscience* 25, no. 3 (2000): 239–40.

Silburn, Lilian. *Kundalini: Energy of the Depths*. Albany: State University of New York Press, 1988.

Speaight, Robert. *The Life of Teilhard de Chardin*. New York: Harper & Row, 1967.

Stapledon, Olaf. *Last and First Men* and *Star Maker*. New York: Dover, 1968.

Stapp, H.P. *The Mindful Universe: Quantum Mechanics and The Participating Observer*. New York: Springer, 2007.

Steiner, Rudolf. *The Evolution of Consciousness as Revealed Through Initiation-Knowledge: Thirteen Lectures Given at Penmaenmawr, North Wales 19th to 31st August, 1923*. Translated by V.E.W. and C.D., 2nd ed. Great Britain: Rudolph Steiner Press, 1966.

———. *How to Know Higher Worlds: A Modern Path of Initiation*. Translated by Christopher Bamford. Anthroposophic Press, 1994.

———. *Knowledge of the Higher Worlds and Its Attainment*. Translated by George Metaxa, 3rd ed. Hudson, NY: Anthroposophic Press, 1947. First published 1904.

———. *An Occult Physiology: Eight Lectures by Rudolf Steiner, Given in Prague, 20th to 28th March, 1911.* 2nd ed. London: Rudolf Steiner Publishing, 1951.

———. *What Is Anthroposophy? Three Perspectives on Self-Knowledge.* Edited by Christopher Bamford. Great Barrington: Anthroposophic Press, 2002.

Stuart, C. I. J. M., Y. Takahashi, and H. Umezawa. "Mixed-System Brain Dynamics: Neural Memory As a Macroscopic Ordered State." *Foundations of Physics* 9, no. 3–4 (1979): 301–27.

Susskind, Leonard. *The Black Hole War: My Battle with Stephen Hawking to Make the World Safe for Quantum Mechanics.* New York: Little, Brown and Company, 2008.

Swami Hariharānanda Āranya. *Yoga Philosophy of Patañjali.* Translated by P. N. Mukerji. Albany: State University of New York Press, 1983.

Swami Lokeswarananda. *Taittiriya Upanishad: Translated and with Notes Based On Sankara's Commentary.* Calcultta: Ramakrishna Mission Institute of Culture, 2010.

Swami Satchidananda, trans. *The Yoga Sutras of Patañjali.* Yogaville, VA: Integral Yoga Publications, 1990.

Taimni, I.K. *Gayatri: The Daily Religious Practice of the Hindus.* Wheaton, IL: Quest Books, 1989. First published 1946.

———. *The Science of Yoga.* India: The Theosophical Publishing House, 1961.

———. *Self-Culture: The Problem of Self-Discovery and Self-Realization in the Light of Occultism*. Madras: The Theosophical Publishing House, 1945.

Teilhard de Chardin, Pierre. *Activation of Energy*. Translated by Rene Hague. London: William Collins Sons, 1976.

———. "The Activation of Human Energy." In *Activation of Energy*, translated by René Hague, 359–93. London: William Collins Sons, 1976. First published 1953.

———. "The Atomism of Spirit." In *Activation of Energy*, translated by René Hague, 21–57. London: William Collins Sons, 1976. First written 1941.

———. "Centrology: An Essay in a Dialectic of Union." In *Activation of Energy*, translated by René Hague, 97–127. London: William Collins Sons, 1976. First written 1944.

———. *Christianity and Evolution: Reflections on Science and Religion*. Translated by René Hague. London: William Collins Sons, 1971.

———. "The Convergence of the Universe." In *Activation of Energy*, translated by René Hague, 281–96. London: William Collins Sons, 1976. First written 1951.

———. “The Death-Barrier and Co-Reflection, or the Imminent Awakening of Human Consciousness to the Sense of Its Irreversibility.” In *Activation of Energy*, translated by René Hague, 395–406. London: William Collins Sons, 1976. First written 1955.

———. 1960. *The Divine Milieu*. New York: Harper & Row.

———. “The Energy of Evolution.” In *Activation of Energy*, translated by René Hague, 359–72. London: William Collins Sons, 1976. First published 1953.

———. “The Formation of the Noosphere.” In *Man’s Place in Nature: The Human Zoological Group*, translated by René Hague, 96–121. New York: Harper & Row, 1956.

———. “The Great Monad.” In *The Heart of Matter*, translated by René Hague, 182–95. New York: Harcourt Brace Jovanovich, 1978. First published 1918.

———. *The Heart of Matter*. Translated by René Hague. New York: Harcourt Brace Jovanovich, 1978.

———. “Hominization.” In *The Vision of the Past*, translated by J. M. Cohen, 51–79. New York: Harper & Row, 1966. First written 1923.

———. *The Human Phenomenon*. Translated and edited by Sarah Appleton-Weber. Portland, OR: Sussex Academic, 2003. First published 1955.

———. *Letters from a Traveler*. New York: Harper & Row, 1962.

———. *Lettres Intimes de Teilhard de Chardin a Auguste Valensin, Bruno de Solages, et Henri de Lubac 1919–1955*. Paris: Aubier Montaigne, 1972.

———. "Life and the Planets." In *The Future of Man*, translated by Norman Denny, 97–123. New York: Harper & Row, 1959. First published 1945.

———. "My Fundamental Vision." In *Toward the Future*, translated by René Hague, 163–208. London: William Collins Sons, 1975. First published 1948.

———. "Nostalgia for the Front." In *The Heart of Matter*, translated by René Hague, 168–81. New York: Harcourt Brace Jovanovich, 1978. First published 1917.

———. *The Phenomenon of Man*. Translated by Bernard Wall. New York: Harper & Row, 1959.

———. *Pierre Teilhard de Chardin: L'Oeuvre Scientifique*. Edited by Nicole and Karl Schmitz-Moormann. 10 vols. Munich: Walter-Verlag, 1971.

———. "A Sequel to the Problem of Human Origins: The Plurality of Inhabited Worlds." In *Christianity and Evolution: Reflections on Science and Religion*, translated by René Hague, 229–36. London: William Collins Sons, 1971.

———. "The Spirit of the Earth." In *Human Energy*, translated by J. M. Cohen, 93–112. New York: Harcourt Brace Jovanovitch, 1969. First written 1931.

———. "Some Notes on the Mystical Sense: An Attempt at Clarification" In *Toward the Future*, translated by René Hague, 209–11. London: William Collins Sons, 1975. First published 1951.

———. "The Zest for Living." In *Activation of Energy*, translated by René Hague, 229–43. London: William Collins Sons, 1976. First written 1950.

Tiller, William. *Science and Human Transformation.* Berkeley, CA: Pavior, 1997.

Vernadsky, Vladimir I. *The Biosphere.* Translated by D.B. Langmuir. New York: Springer-Verlag, 1998.

Walker, J. Samuel. *Three Mile Island: A Nuclear Crisis in Historical Perspective.* Berkeley: University of California Press, 2004.

Wallace, B. Alan. *Fathoming the Mind: Inquiry and Insight in Fathoming the Mind: Inquiry and Insight in Düdjom Lingpa's Vajra Essence.* Somerville, MA: Widsom Publications, 2018.

———. "Introduction." In *Heart of the Great Perfection* by Düdjom Lingpa, translated by A. B. Wallace, 1–26. Somerville, MA: Wisdom Publications, 2015.

Ward, Benedicta, ed. *The Desert Fathers: Sayings of the Early Christian Monks.* London: Penguin Books, 2003.

Whicher, Ian. *The Integrity of the Yoga Darsana: A Reconsideration of Classical Yoga.* New York: State University of New York Press, 1998.

Wiener, Norbert. *Cybernetics: Or Control and Communication in the Animal and the Machine.* Cambridge: MIT Press, 1948.

Wilber, Ken, ed. *The Holographic Paradigm and Other Paradoxes*. Boulder: Shambhala, 1982.

Wiltschko, Wolfgang, and Roswitha Wiltschko. "Magnetic Orientation and Magnetoreception in Birds and Other Animals." Journal of Comparative Physiology A: Neuroethology, Sensory, Neural, and Behavioral Physiology 191 (2005): 675–93.

Wood, Ernest and Paul Brunton. *Practical Yoga, Ancient and Modern: Being a New, Independent Translation of Patanjali's Yoga Aphorisms, Interpreted in the Light of Ancient and Modern Psychological Knowledge and Practical Experience.* Chatsworth, CA: Wilshire Book Co., 1976.

Woodroffe, Sir John. "Foreward." In Vasant Rele, *The Mysterious Kundalini: The Physical Basis of the 'Kundalini Yoga,'* ix–xi. Bombay: D.B. Taraporevala Sons & Co., 1927.

Woodroffe, Sir John. *The Serpent Power: Being the Ṣaṭ-cakra-nirūpana and Pādukā-pañcaka*. Reprint ed. New York, Dover, 1974. First published 1918.

Woodroffe, Sir John. *The World As Power*. 3rd ed. Madras: Ganesh & Co, 1966.

Yirka, Bob. "New Study Strengthens Olfactory Vibration-Sensing Theory." *Phys.org*, January 29. 2013.

https://phys.org/news/2013-01-olfactory-vibration-sensing-theory.html.

Zambito, Salvatore. *The Unadorned Thread of Yoga: The Yoga Sutra of Patañjali in English - A Compilation of English Translations of the Yoga Sutra of Patañjali.* Poulsbo, WA: The Yoga Sutras Institute Press, 1992.

Endnotes

---

[1] Nikola Tesla (2019) “The Law of Reflection.” by A.G. Venera, Novum Publishing. P. 6

[2] Pribram, *The Form Within*, 526.

[3] Graphic by Darth Kule (2010). Public domain. Image retrieved from Wikimedia Commons. https://commons.wikimedia.org/wiki/File:Black_body.svg

[4] Bohm, *Wholeness and the Implicate Order,* 188.

[5] Bohm, 189.

[6] Bohm, 189.

[7] Graphic by Camille Flammarion (Paris, 1888). Reprinted under the terms of a Creative Commons Attribution ShareAlike 3.0 Unported license. Image retrieved from Wikimedia Commons.

[8] Graphic by J. Wheeler (Austin, 1990). Reprinted under the terms of a Creative Commons Attribution ShareAlike 3.0 Unported license. Image retrieved from Wikimedia Commons.

[9] Ibid., 142.

[10] Pribram, “What the Fuss Is All About,” 29.

[11] Pribram, “Prolegomenon for a Holonomic Brain Theory.”

[12] Pribram, “Brain and Mathematics,” 232.

[13] Pribram, *The Form Within*, 495.

[14] Kuo, *Network Analysis and Synthesis*, 1.

[15] Pribram, *Brain and Perception*, 73.

[16] Stein and Shakarchi, *Fourier Analysis*, 134–36.

[17] Kuo, *Network Analysis and Synthesis*, 40.

[18] Wiener, *Cybernetics: Control and Communication in Animal and Machine*, 198.

[19] Ibid.

[20] Ibid., 202.

[21] Ibid. Italics added.

[22] Mandelbrot, "Fractals and the Rebirth of Iteration Theory," 151.

[23] Penrose, *The Emperor's New Mind: Concerning Computers, Minds and the Laws of Physics*, 124.

[24] Ibid., 127.

[25] Jung, "On the Nature of the Psyche," 215.

[26] Kuo, *Network Analysis and Synthesis*, 13.

[27] Ibid.

[28] Ibid.

[29] Ibid., 3.

[30] Bohm, *Quantum Theory*, 1.

[31] McGilchrist, *The Matter with Things*.

[32] Krishnamurti and Bohm, "On Intelligence."

[33] Grof, "Revision and Re-Enchantument of Psychology," 137.

[34] Bohm, *Wholeness and the Implicate Order*, 291.

[35] Fechner, Religion of a Scientist, 158.

Made in the USA
Monee, IL
04 November 2026

67375816R00075